EXERCISES IN
PILOTAGE

EXERCISES IN
PILOTAGE

A Basic Introduction with Tests and Answers

JOHN ANDERSON

Illustrated by A. J. P. Anderson

David & Charles
Newton Abbot   London   North Pomfret (Vt)

For Jane

British Library Cataloguing in Publication Data

Anderson, John
  Exercises in pilotage: a basic introduction
  with tests and answers.
  1. Pilots and pilotage – Problems,
  exercises, etc.   2. Boats and boating
  I. Title
  623.89′22′076   VK1661

  ISBN 0–7153–8899–1

Photoset by
Northern Phototypesetting Co, Bolton
Printed in Great Britain
by A. Wheaton & Co Ltd, Exeter
for David & Charles Publishers plc
Brunel House Newton Abbot Devon

Published in the United States of America
by David & Charles Inc
North Pomfret Vermont 05053 USA

Contents

Preface

Every summer an armada of about half a million recreational sailors takes to the waters surrounding these islands. The vast majority know what they are about but a small percentage, either through carelessness, ignorance or inexperience, become a statistic in the records of the search and rescue services. Nearly all of these casualties occur within sight of land.

Whenever a vessel is closing the coast the skipper must keep a closer watch than usual and take every precaution to ensure the safety of his craft. This requires the ability to assess the vessel's position with speed and accuracy and to forecast where it will be in the immediate future.

This book has been designed to give beginners an introduction to pilotage. The skills and techniques of navigation may only be learnt by practice at sea, but the author hopes that this volume will give the beginner the basic confidence to commence that practice. Constant practice is essential and to that end plenty of exercises have been given. All the answers are at the back of the book.

Penzance,
1986.

1 Charts and Instruments

Navigation is an art and a science. The navigator uses scientific principles, but in selecting the information to use he exercises judgement or skill, and that is art. It consists of finding your vessel's position and conducting her safely from there to her destination. If that statement smacks of the blindingly obvious, then ponder a second. Suppose you have a navigational problem. Do you know where you are? If the answer is 'Yes' then plot a course clear of all dangers. If 'No', then fix your position by the best means possible and proceed as above. Your problem is solved.

Pilotage is the navigation of coastal waters within sight of land, rock-lighthouses, light-vessels and buoys. For our purposes, it will also include cross-channel passages, when one is temporarily out of sight of land. The aim of this book is to tell you most of what you need to know in order to take a small vessel to sea with confidence and in safety.

The Tools of the Trade

First and foremost you must purchase charts and for the purpose of this book's exercises you will need Admiralty Charts 5067 and 5051, Trevose Head to Dodman Point, and Land's End to Falmouth respectively. The reason for this choice is that these two represent one of the most interesting and dangerous stretches of English coastline, possessing as it does no less than twelve lighthouses and one lightvessel. These are instructional charts, being cheaply printed on thin paper and must never be used at sea because they lack the latest corrections. For use at sea you must buy their more expensive counterparts, 2565 and 777, and for passage planning, 2675 The English Channel. All these can be bought at yacht chandlers.

If you live inland and far from a chandler the following firms provide a mail order service.

Captain O. M. Watts, Albermarle St, London W1
Kelvin Hughes   31 Mansell St, London E1
                 21 Regent Quay, Aberdeen AB1 2AH
                 3 Central Rd, Eastern Docks, Southampton
                 SO1 2AH
                 375 W. George St, Glasgow G2 4LR

Charts can also be bought direct from the Hydrographer of the Navy, Taunton.

Next, you need a parallel ruler. Your best bet is probably a 12 inch perspex rule (Captain Field's), but in a really small boat you may find a Douglas or Portland Protractor more handy.

A sophisticated version of both is a Hurst Plotter. They are all in effect movable compass roses and provided they have been lined up accurately with latitude and longitude, are just as effective as a parallel rule. Each comes with full instructions for its use. For a larger boat with a roomy chart table you may choose to indulge in an 18 inch roller rule, which is more expensive.

Third, you need dividers. Sea-going dividers need to be heavy, non-magnetic and corrosion-proof, and have a span of twelve inches. Their weight prevents them from sliding about in a seaway. It is better not to rely on cheap school dividers even for practice.

Your pencils should be soft and so should your eraser. A 2B pencil is ideal because a harder pencil indents the chart and renders it unusable all the quicker. In fact, you are recommended to buy two copies each of 5067 and 5051 as you will find that even with care they will quickly deteriorate while performing the exercises in this book. You will also need a pair of compasses for describing arcs.

Lastly an Almanac, such as Reed's or Macmillan's Silk Cut is essential. It doesn't have to be the current edition. One can sometimes obtain the previous year's edition cheaply from public libraries; but it shouldn't be more than two years old. If you choose this option you will still have to purchase the times of high and low water for the current year for Dover, or, for the charts I have listed, Devonport.

Finding Your Way About the Chart

You will, having purchased your charts, immediately notice that one is black and white and the other coloured. The coloured one is metric and all depths, or soundings, are in metres, as are heights of lights and high points on land. Soundings on black and white charts are in fathoms, and heights are in feet.

ALWAYS check the units being used for soundings which are clearly stated below the title of the chart. There are yachts which would still be afloat today if someone hadn't assumed they had three fathoms beneath the keel when in fact they had only three metres. Later on you will come across harbour plans where the soundings may be in feet.

You will also notice that, unlike Ordnance Survey maps, your charts are liberally scattered with numbers. It is important to remember that these, if soundings, are the depths of water below Chart Datum, which is theoretically the least amount of water you are ever likely to experience. Heights, however, are measured from Mean High Water Springs (MHWS), in theory the most amount of water. Theoretically, because sea levels can be affected by barometric pressure, storm surges and on or offshore winds.

To really understand all the information on a chart it is necessary to buy Admiralty Publication 5011, Symbols and Abbreviations. It is not necessary to learn all the symbols, but those shown here are vital and should be committed to memory. **Fig 1.1.**

Then there are the symbols which we shall add to the chart. It is vital to learn this notation, and that everyone keeps to it. Imagine how confused a crew would become if everyone using the chart employed their own private form of shorthand. **Fig 1.2**

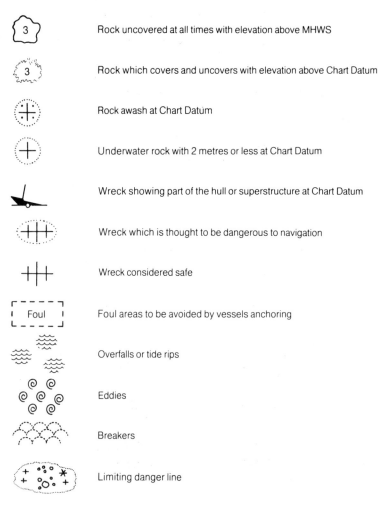

Rock uncovered at all times with elevation above MHWS

Rock which covers and uncovers with elevation above Chart Datum

Rock awash at Chart Datum

Underwater rock with 2 metres or less at Chart Datum

Wreck showing part of the hull or superstructure at Chart Datum

Wreck which is thought to be dangerous to navigation

Wreck considered safe

Foul areas to be avoided by vessels anchoring

Overfalls or tide rips

Eddies

Breakers

Limiting danger line

PA Position Approximate   PD Position Doubtful   ED Existence Doubtful

Fig 1.1   Chart symbols

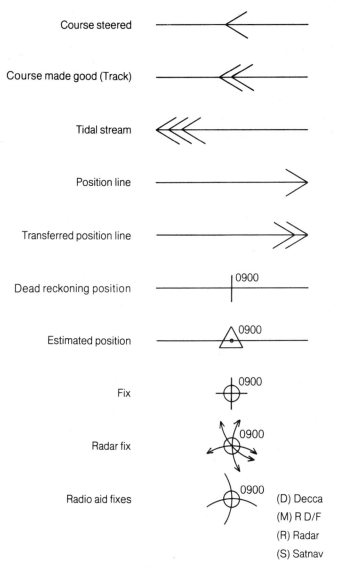

Course steered

Course made good (Track)

Tidal stream

Position line

Transferred position line

Dead reckoning position   0900

Estimated position   0900

Fix   0900

Radar fix   0900

Radio aid fixes   0900

(D) Decca
(M) R D/F
(R) Radar
(S) Satnav

Fig 1.2 Symbols to use on the charts

EXERCISE ONE

(a)   How does pilotage differ from navigation?

(b)   Why must an instructional chart never be used for navigation?

(c)   What qualities must a boat's dividers possess?

(d)   Why is a 4H pencil unsuitable for chartwork?

(e)   How can you instantly recognise a metric chart?

(f)   Name the three units used on charts for soundings.

(g)   Which measurements are taken from Chart Datum and which from MHWS?

(h)   On a chart, how does a dangerous wreck differ from one considered safe?

(i)   Some rocks are covered at certain times. How is this shown on a chart?

(j)   Which Admiralty publication provides a full list of symbols?

(k)   Make drawings of the symbols for courses, both steered and made good, and for Dead Reckoning, Estimated and Fixed Positions.

2 Latitude and Longitude

When in sight of land it is common practice to give the ship's position relative to a particular object, usually as a range and bearing from it. For example, one might say 'I am two miles due south of Lizard Point'. Out in mid-channel, however, or in poor visibility, we need to be able to give our position relative to a fixed grid which we imagine to be on the surface of the earth.

Latitude is the angular distance of a place north or south of the equator. Lines of latitude are drawn across the chart and latitude is measured on the sides of the chart in degrees and minutes. One degree contains sixty minutes, and since the charts we are using are situated in the Northern Hemisphere, latitude readings are suffixed by North. **Fig 2.1**

Using your dividers, place one point on the position whose latitude you wish to find and open the dividers so that the other arm rests on the nearest parallel of latitude due north or south of it. Transfer the dividers to the side of the chart nearer the object, and keeping one arm on the parallel chosen, read off

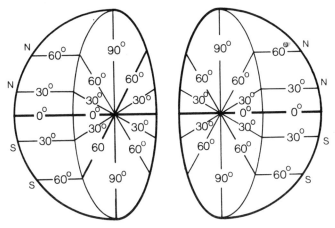

Fig 2.1 Latitude

the degrees and minutes indicated by the other arm. **Fig 2.2** will make this clear.

Suppose you require the latitude of the object at A. Having transferred to the latitude scale you read off 50 degrees 01 minutes. In the case of object B the answer is something less than 50 degrees. It is two minutes less and since there are sixty minutes in a degree the answer is 49 degrees 58 minutes.

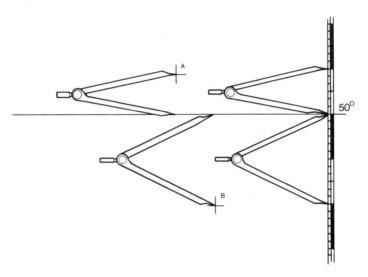

Fig 2.2   Use of dividers to measure latitude

In the following exercises you are required to find the latitude of various features on the chart. The information in brackets is simply to help you find them.

EXERCISE TWO
Using chart 5067 find the latitude of

(a) Manacle Rocks buoy (7 cms south of Falmouth)
(b) St Michael's Mt (3 cms east of Penzance)
(c) St Anthony Hd Lt (Eastern entrance to Falmouth)
(d) Lizard Pt Lt (most southerly point on mainland)
(e) Wolf Rock Lt (10 cms SW of Land's End)

Longitude

Longitude is the angular distance east and west of the Prime or Greenwich Meridian. Lines of longitude are drawn vertically from top to bottom of the chart, and although parallel for all practical chartwork purposes, naturally converge at the poles. We measure off the longitude along the top and bottom of the chart whichever is nearer, in just the same way as you measured off latitude. When you are anywhere west of Greenwich the figure is suffixed with a W. The figures are measured from the prime meridian and so increase as you move east or west from it.

EXERCISE THREE
Give the longitude of

(a) Longships Lt House (1.5 cms west of Land's End)
(b) Pendeen Lt House (7.5 cms NNE Land's End)
(c) Trevose Hd Lt (the most northerly light)
(d) Seven Stones Light Vessel (17 cms west of Land's End)
(e) Bishop Rock Lt (the most southerly light in the Isles of Scilly)

Now let us combine latitude and longitude and this gives us a grid reference by which we can pinpoint exactly any point on the chart.

For instance, the position of Round Island Lt House (the most northerly of the Scillies lights) is 49° 58.8′ N, 06° 19.2′ W. The decimal points signify tenths of a minute. You will have noticed that each minute is divided to represent .2, .4, .6, and .8 of a minute respectively. If your divider's point falls between two divisions interpolate .1, .3, .5, or .9 as the case may be.

To complete the next exercise you will probably find that dividers alone are insufficient and the parallel rule will be needed. Examine the top or bottom scales of the chart to identify the area of longitude you need. For example, with Round Island we need the bottom left-hand corner of the chart. Place one edge of the ruler on the light symbol and keeping it parallel with the nearest parallel of latitude, read off its latitude from the left-hand scale.

Conversely in Exercise Four (a) you should be looking at the right-hand corners (05° 01′ W). Lining up the ruler with the nearest latitude (which might well be the top edge) 'walk' it down to 50° 32.9′ N and lightly draw a line across the sea area and coast. Place the ruler on one side and take up the dividers. With one point on 05 degrees West stretch out 02 minutes and repeat the process on the line you have just drawn. The point should be on, or very close to, a lighthouse.

EXERCISE FOUR
Name the feature shown on the chart at

(a) 50° 32.9′ N   05° 03′ W.
(b) 50° 26′ N   05° 10.9′ W.
(c) 50° 20′ N   05° 13.9′ W.
(d) 49° 58′ N   05° 17.5′ W.

3 Distances and Bearings

In Chapter Two you were given distances of objects in centimetres to help you in their location, but naturally, since charts come in many different scales, this will not do, and we must know the exact distances that objects are apart. Fortunately, this is simplicity itself.

It so happens that the LATITUDE scale on the SIDE of the chart serves a dual purpose. Each minute of Latitude (never Longitude unless you are on the Equator) represents one nautical mile, or 6,080 feet. Let us suppose that you need to know the distance from Lizard Point to Tater-du Lt (SW of Newlyn). Using your dividers place one point on each feature and transfer to the side of the chart in approximately the same latitude as the objects. If you place one point on 50° 00′ N you merely count the minutes (ie miles) to the other point. In this case it is 15.4 n miles.

As we have already seen each division of the minute marking represents ²⁄₁₀ of a mile. Since for all practical purposes one nautical mile equals 6,000 feet, it represents $\dfrac{2 \times 6{,}000 \text{ ft}}{10}$ or 1,200 ft or 400 yards. This is useful because another measurement of distance at sea is a cable, which is 200 yards or ¹⁄₁₀ of a nautical mile. These divisions therefore represent 2, 4, 6 and 8 cables which can be interpolated for odd numbers.

It is customary when giving a range at sea to use decimal points if over one mile, but cables under, eg 2.9 miles but 7 cables.

EXERCISE FIVE
What are the straight line distances between

(a) Manacle Rks buoy and Dodman Pt (on right hand edge)?
(b) Penlee Pt buoy and Mountamopus buoy (both Mount's Bay)?

(c) Longships Lt Ho and Wolf Rock Lt Ho?
(d) Lizard Pt Lt Ho and Wolf Rock Lt Ho?
(e) Pendeen Lt Ho and Seven Stones Lt vessel?

Bearings

Directions between objects at sea are measured in degrees, clockwise from North. In days of yore it was necessary to be able to 'box the compass' by calling all thirty-two points and a favourite punishment for an inattentive student at Sea Training School was to be made to recite the eight points from south to east, backwards! For any sadists who are interested, this mild torture consisted of: South, South by East, South-South-East, South-East by South, South-East, South-East by East, East-South-East, East by South and East. Fortunately, this chore is now a thing of the past, and the compass card is divided much more accurately into 360 degrees. It is still

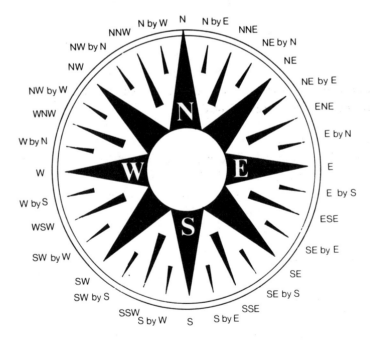

Fig 3.1   Points of the compass

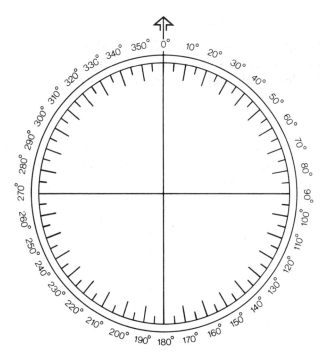

Fig 3.2 The notational compass

important though, to relate degrees to the eight major points of the compass. **Figs 3.1 and 3.2**

EXERCISE SIX
Complete this table, remembering always to use three-figure notation (eg East is 090° NOT 90°)
(a) North East   =   degrees
(b)   = 315 degrees
(c) 202½ degrees   =
(d)   = 337½ degrees
(e) East North East   =

To find a bearing on the chart use the parallel ruler. Place one edge so that it passes through two points required and keeping that arm absolutely still 'walk' the other to its fullest extent. If it passes through the centre of a compass rose, all

well and good. If it does not, hold that arm steady while the
first is brought up to it. Repeat as many times as necessary until
an edge crosses the centre spot (this takes a little practice).
Remember, the less moves made the greater the accuracy.

Once the ruler is over the centre of the rose, the edge will
pass through two bearings opposite each other. Decide which
one you want, as the other is its reciprocal.

A RECIPROCAL is the opposite bearing or course to that
required and thus is 180 degrees out. Soon after using the
compass rose you will read off a reciprocal in error. Welcome
to the Club! It numbers many millions. It is easier to make this
mistake with three-figure notation than with quadrantal
notation. After all, one would hardly say North when one
meant South, but it is easy to read off 270° when you mean
090°.

EXERCISE SEVEN
What are the reciprocals of

(a) 000°   (b) 315°   (c) 090°   (d) 180°   (e) 227°?

In the next exercise remember that your starting point is the
second place and you are looking towards the first. Use the
OUTER Compass Rose.

EXERCISE EIGHT
What is the TRUE bearing of

(a) Pendeen Lt FROM Godrevy Lt (East of St Ives Bay)?
(b) Runnelstone Buoy FROM Lizard Lt?
(c) Lizard Lt FROM Wolf Rock Lt?
(d) 49° 55′ N, 04° 57′ W FROM 49° 47′ N, 05° 14′ W?
(e) 50° 21′ N, 06° 04′ W FROM 50° 05′ N, 05° 55′ W?

In Chapter Two I said that positions were often given
relative to a fixed object, and this is usually a range and
bearing. For example, 186° FROM St Anthony Hd range 5.7
miles.

EXERCISE NINE

What is at

(a) 201° Cape Cornwall 4 miles?
(b) 248° Pendeen Lt 16.6 miles?
(c) 280° Lizard Pt Lt 3.2 miles?
(d) 204° Land's End 8.1 miles?
(e) 173° St Anthony Hd Lt 8.2 miles?

4 Time, Speed and Distance

To be able to estimate your position and forecast any future position or time of arrival (ETA) it is essential to know your speed through the water. Those last three words are important because the surface of the sea is constantly moving and your speed OVER THE SEABED is vital to your assessment of your position relative to the land. If you know your speed through the water, then by adding or subtracting known tides or leeway due to wind, you can obtain a reasonably accurate estimation of your speed over the ground, or speed made good.

You can calculate it in several ways. A modern yacht or cruiser is likely to have an impeller log projecting about three inches from the hull, which, spinning as you travel through the water, will activate a speedometer or odometer in the cockpit. By checking the distance travelled against time you will obtain an average speed which is the first step in estimating your position. **Fig 4.1**

second, you may have a patent log. This consists of a brass rotator attached to a sinker or 'fish' trailed astern on the end of a long braided line. At the inboard end, secured to the transom or pushpit, is a register which records distance travelled. Although accurate, the patent log has its disadvantages in that it must be streamed only when well clear of harbour and

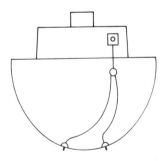 

Fig 4.1   Impeller log

retrieved before entering busy waters. Retrieval consists of slowing down and paying out the line from the inboard end as you haul in the rotator, to prevent kinks from forming. Once inboard, the rotator, fish and line must be washed in FRESH water and the line hung to dry. The brass parts must be lightly oiled and stowed away. **Fig 4.2**

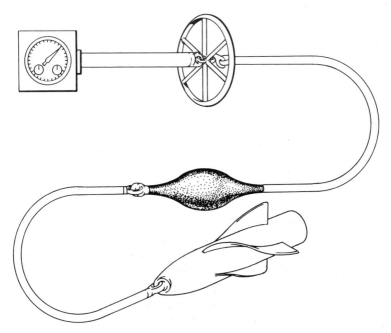

Fig 4.2   Patent log

Third, a home-made device may be used for short intervals. It consists of a triangular piece of wood, about thirty centimetres across, with three holes drilled in it. A strip of lead is crimped to the base while two ends of a log line are passed through the lower holes. A third end is secured to a peg which is driven into the upper hole. After a length of twenty fathoms (120 feet) is paid out to keep the log clear of your wake, knots or short lengths of white bunting are tied in the line at regular intervals, say ten feet apart. Having cast the float over the stern and let the stray line out, start a stopwatch as the first knot goes over the side. Five knots later, or fifty feet on, stop the watch and secure the line.

Let us say that the stopwatch registers four seconds, then your speed is $^{50}\!/\!4 = 12.5$ ft/sec which is 7.5 knots. This is obtained by multiplying the ft/sec figure by 0.6 which is the relationship of feet in a nautical mile to the number of seconds in an hour.

Finally, tug the line, pulling the peg out of the board's apex, and haul in. **Fig 4.3**

Fig 4.3   Home-made log

Lastly, in a small sailboat or dinghy, provided you know the length of your craft and possess a few odd bits of wood or even a cigarette packet, and a stopwatch, you can use the simplest method of all, the Dutchman's Log.

Dutchman's Log

This is an invaluable method of finding your speed when it is so low as to render an impeller or patent log inaccurate. Send an assistant into the bows with a small buoyant object such as a piece of wood and instruct him to throw it ahead and clear of your wash. Tell him to signal when it passes the bow and start your watch. In the stern you will sight it and stop the watch when it passes you. Having decided whether you will use the overall or waterline length of your boat, you can draw up a table by multiplying the time interval by the factor of 0.6 and the answer will be your speed in knots.

Say, for example, your craft is 30 feet long and the 'log' took 1 second to pass your speed would be $30 \times 0.6 = 18$ knots.

DUTCHMAN'S LOG TABLE (for a 30 foot vessel)

Interval (seconds)	Speed (knots)	Interval	Speed
2	9.0	5	3.6
3	6.0	6	3.0
4	4.5	7	2.5

and so on.

Having discovered your speed you can now perform some simple calculations with regard to time and distance, as these three units are all functions of each other. For example, $\frac{\text{distance}}{\text{time}} = \text{Speed}$ as in $\frac{12 \text{ miles}}{3 \text{ hours}} = 4 \text{ knots}.$

Always remember, a knot is a unit of speed, not distance. In other words 4 knots = 4 nautical miles per hour. A common mistake is to refer to so many knots per hour, which would be a unit of acceleration.

A handy device is the Speed, Time and Distance Triangle. By covering up the unit required one is provided with the formula for calculating it. **Fig 4.4**

  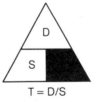 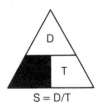

D = S × T   T = D/S   S = D/T

Fig 4.4   Speed, time and distance triangle

EXERCISE TEN

Work out the speed if you covered
(a) 15 miles in 3 hours
(b) 10 miles in 2 hours
(c) 11 miles in 4 hours
(d) 9 miles in 2.5 hours
(e) 7.5 miles in 1.5 hours

Work out the distance covered if you travelled for

(f) 4 hours at 3 knots
(g) 3 hours at 7 knots
(h) 2.5 hours at 4 knots
(i) 1.25 hours at 6 knots
(j) 7.5 hours at 4.5 knots

EXERCISE ELEVEN (use chart 5067)
(a) You are in position 50° 20′ N, 06° 00′ W. You steer 190° for 1 hour at 16.5 knots. Where do you end up?
(b) You are in position 50° 22.8′ N, 06° 06.7′ W. One hour later you are in position 50° 01.2′ N, 06° 00.8′ W. What has been your course and speed?
(c) You are in position 50° 20′ N, 05° 20′ W. You steer 300° at 10 knots. One hour later you are in position 50° 24′ N, 05° 35.2′ W. What has been your ACTUAL course and speed?

5 Position Finding

You have now reached a point where you can fulfil the first requirement of a navigator, stated at the beginning of this book, namely to find your position.

There are various methods of obtaining this information and in ascending degrees of accuracy they are

DEAD RECKONING (DR) position using course and speed only. Dead has nothing to do with a state of non-existence, but refers to Ded, an abbreviation of Deduced Reckoning. **Fig 5.1**
ESTIMATED POSITION (EP) is somewhat more accurate, as it takes into account tidal streams and leeway. **Fig 5.2**

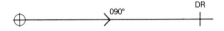

Fig 5.1   Dead reckoning (DR) position

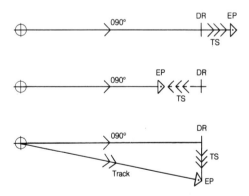

Fig 5.2   Three estimated positions (EP)

FIX This is the only reliable method of finding position and there are many types available to the mariner. The first is Celestial, using the sun, moon, stars and planets. It is not within the scope of a book on pilotage.

Second, and the one on which we shall concentrate, is the Terrestial Fix using land-based objects.

Third is the Radar Fix which will be dealt with in some detail in Chapter 13. Lastly, there is the Radio Fix which involves Radio Direction Finding Beacons, Decca chains and Satellite Navigation, covered in Chapter 14.

Terrestrial Fix

The most common type of fix is a cross bearing fix. Let us assume that Lizard Pt Lt bears 270° from your position. Then

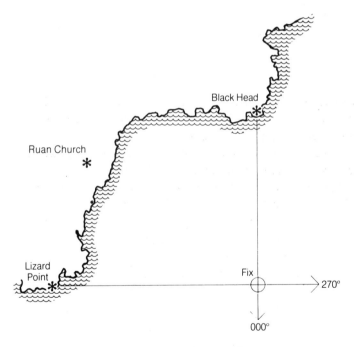

Fig 5.3   Cross-bearing fix

it is due west of you and you are due east of it. A line drawn out into Falmouth Bay at 270° (or 090°) is your position line AND YOU ARE SOMEWHERE ON THAT LINE. That is all you know at present. But if you now take a bearing on Black Head and find it bears 000° then you can also say that you are somewhere on that line. Theoretically your position is where the two cross. **Fig 5.3**

EXERCISE TWELVE (Chart 5067)
What will you find where

(a) Longships Lt bears 019° and Seven Stones Light Vessel (LV) bears 305°?
(b) Lizard Pt Lt bears 252° and Manacles buoy bears 328°?
(c) Trevose Hd Lt bears 039° and Towan Hd (NW of Newquay) bears 102°?
(d) Gurnard's Hd (W of St Ives) bears 142° and St Agnes Hd bears 091°?
(e) Lizard Pt Lt bears 268° and Dodman Pt bears 355°?

Transits

An extremely useful, and somewhat underestimated method of fixing your position, which should be used whenever opportunity permits, is the transit.

The transit is familiar to anyone who has ever batted in or umpired a cricket match, because a transit is what is asked for and given whenever a fresh batsman walks to the crease and asks for 'Centre'. The umpire sights over his middle stump and directs the bat, held vertically, until it is line (or transit) with the batsman's stump. Similarly, if at sea two *charted* objects come into line and you plot that line, then you are somewhere on it. If you are fortunate enough to sight two transits, practically simultaneously, then you have a fix, with no compass error.

Transits are invaluable as leading marks when negotiating narrow, tricky channels and when coming to anchor. It is also advisable to check your compass (or at least the hand-bearing

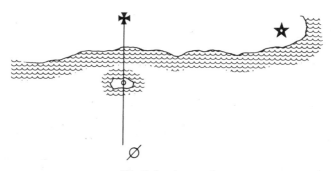

Fig 5.4   A transit

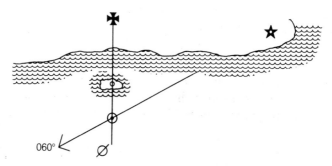

Fig 5.5   Transit and bearing

compass) on a transit before putting to sea. **Fig 5.4**

Now it is not often that you are going to be fortunate enough to have two transits simultaneously, but if you have a transit and a compass bearing on a separate object then you have a perfectly acceptable fix, probably more accurate than a cross-bearing fix as compass error is reduced. But beware. Make sure that the objects you sight are indeed those shown on the chart. Many a yachtsman has come to grief because he was using the wrong White House (conspic). **Fig 5.5**

EXERCISE THIRTEEN (Chart 5067)
(a) Cape Cornwall bore 056°, Longships Lt bore 142°.
(b) Wolf Rk bore 170°, Longships Lt bore 074°, Seven Stones LV bore 275°.
(c) St Michael's Mt bore 015°, range 5 miles.
    What are the latitude and longitude of these fixes?

Three Bearing Fixes

You may have discovered when you attempted Exercise 13(b) that the three position lines failed to cross (or cut) in precisely the same place. At sea they rarely do. This is due to several factors among which are inaccuracy of bearings, wrong objects, compass error, and most common, an acute angle of cut of the position lines. The ideal angles to look for are between 45 and 90 degrees, but this is not always possible.

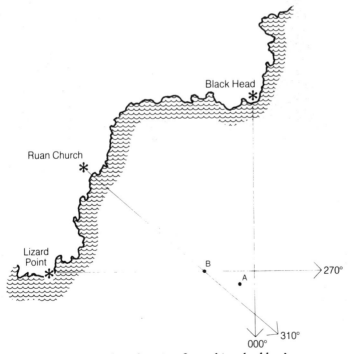

Fig 5.6 Three bearing fix and 'cocked hat'

When, due to poor visibility or lack of objects, an acute angle of cut is the only option then one must make the best of it.

Since the position lines fail to cut at the same point a triangle is formed. Provided that the triangle is small, one may assume that one is in the middle of the triangle and proceed from there (point A in **Fig 5.6**).

However, if you are close in, say within a mile of the coast, and your triangle is one hundred yards (½ cable) or less along its greatest side, it is better to assume a position in the triangle which puts you closest to the nearest danger (point B in **Fig 5.6**). In these circumstances proceed with caution and take another fix as soon as possible. Anything greater than ½ cable, immediately try again, using different objects if at all possible.

Several miles offshore you can be quite happy with a half cable triangle. These triangles, incidentally, are known as 'cocked hats', which tells you how long sailors have been drawing them.

The Running Fix

Supposing there is only one object in sight on which to take a bearing so that you are unable to get crossed position lines. In this case you can use a running fix. Let us assume that at 0900 you are steering 090° at 5 knots and you sight Lizard Pt Lt bearing 045°. Draw in the position line, remembering that at 0900 you are somewhere on that line. Assume your vessel at a point anywhere on that line. From there plot your course (090° for 5 miles) and mark your DR position. This is your DR for 1000. Using your parallel rule transfer the first position line to pass through the 1000 DR position. In theory you will be somewhere on that line at 1000. Place two arrows on each end to show that it is a transferred position line. Lastly at 1000 take a second bearing of the light. It now bears 300° and at 1000 you are therefore somewhere on THAT line. Since you are also theoretically on the transferred position line you must be where they intersect. **Fig 5.7**

The principle behind the running fix is that if you lined up an infinite number of vessels on the first position line and they all

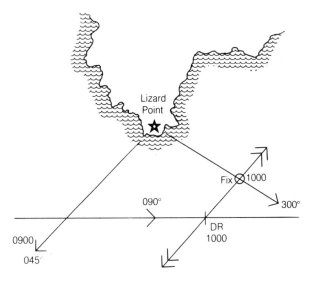

Fig 5.7 A running fix

steamed 090° for one hour at 5 knots, they would all line up on a position line 5 miles east of, and parallel to, the first line, so it matters not from which point on the first line you start your course.

The running fix is not to be preferred because you do not know your position until an appreciable amount of time has elapsed after the initial bearing and it tends to be less accurate than other methods because of miscalculations of speed, tidal stream and leeway, but it is better than nothing when only one object is in sight.

EXERCISE FOURTEEN
Plot your position with a running fix in the following

(a) Godrevy Lt bears 120°. Course 060°, speed 10 knots. Time 1400. At 1430 Godrevy Lt bears 180°. (Remember to halve your speed since only 30 minutes has elapsed between bearings.)

(b) 1200 Lizard Pt Lt bears 030°. Course 086°, speed 10 knots. At 1215 Lizard Pt Lt bears 320°. (Remember to quarter your speed.)

(c) 1330 Wolf Rk bears 215°. Course 255°, speed 12 knots.
     1400 Wolf Rk bears 125°.

Range and Bearing Fix

This is one of the simplest fixes to take but it has been left till
last because it needs a second piece of equipment (other than
the compass) to complete it. A compass bearing of an object is
taken, then a range or distance off along that position line
from the object is plotted. There are four ways to obtain that
range. **Fig 5.8**

First, you can take a vertical sextant angle of the object using
tables found in the almanac. This involves the purchase of a
sextant. Many books describe the use of the sextant in pilotage
for fixing by horizontal angle as well, but it is difficult to
achieve a satisfactory sextant angle in a small vessel in a
seaway. Additionally, with a horizontal sextant angle fix you

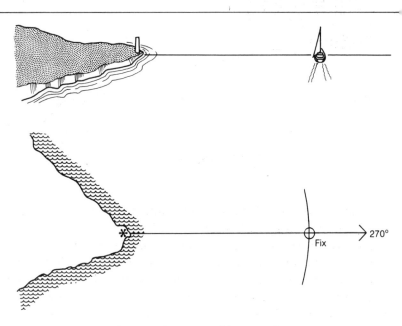

Fig 5.8   A range and bearing fix

have to go below to the chart table and perform some fairly complex geometry. I cannot recommend this method for the beginner.

Second, there is radar, which is dealt with in Chapter 13.

Third, there is the use of the cotangent tables. For under £20 you can purchase a Suunto pocket compass. These are very accurate, well-damped and will provide a useful back-up to your hand-bearing compass. On its base is a table of cotangents which can be used to find ranges. Here's how.

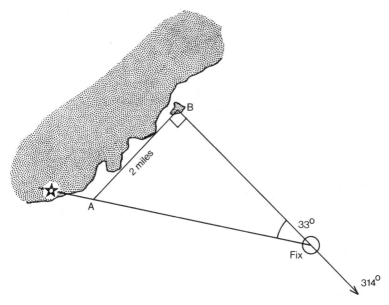

Fig 5.9   The cotangent fix

Let us assume that the angle subtended by two charted objects is 33 degrees. One bears 281° and the other 314°. Plot the two bearings and drop a perpendicular from the nearer point to the other bearing, creating a right angle. Call this line AB and measure it. Let us assume it is two miles. Turn the compass over and read off the cotangent of 33 degrees which is 1.54. Multiply 1.54 by 2 miles and the answer is 3.08 miles which is your distance off. **Fig 5.9**

Lastly, you may construct a DIY rangefinder for a few pence. You need

A card or plastic packing tube about 80 mm (3 inches) in diameter.
Two flat pieces of card and some adhesive tape.
Three short pieces of cotton.

Cut the tube to exactly 420 mm in length or, if you prefer, 15 inches. Cut from the flat card two ends, making a pinhole in the centre of one and a window in the other 25 mm or 1 inch high.

Across the window tape or glue three strands of cotton 7 mm apart, or if you still haven't gone metric, ¼ inch. Now the ratio of each segment to the rangefinder's length is $\frac{7 \text{ mm}}{420 \text{ mm}}$ or $\frac{1}{60}$ or $\frac{¼}{15}$ in.

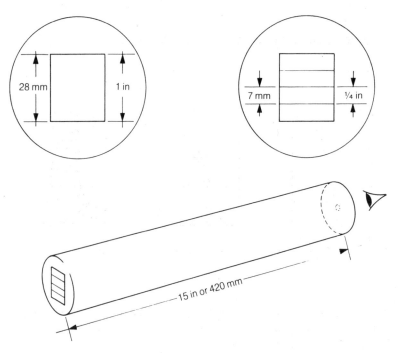

Fig 5.10   Home-made rangefinder

Tape the segmented end over one open end of the tube, and the pinholed end over the other and there is your rangefinder. Shellac it for weather protection and it could last for years. If you are of a practical turn of mind you could make something grander like a student of mine who made his up in brass, beautifully brazed and soldered. **Fig 5.10**

The rangefinder works like this. Suppose you wish to know your distance off the Longships Light. The chart tells us it is 34 metres above MHWS but its elevation is probably greater because it is not yet high water. Assume it to be half-tide and add 3 metres to the height, ie 37 metres. Now sight through your rangefinder and you find the light occupies one segment. Multiply 37 metres by 60 and that is your distance off.

$$37 \text{ m} \times 60 = 2220 \text{ metres} = 2.2 \text{ km} = 1.37 \text{ miles}$$

Should the light occupy two segments then it is obviously closer and the distance will be 1.1 km or 6.8 cables, while three segments occupied is $\frac{2.2 \text{ km}}{3} = 0.73$ km or 4.5 cables. Four segments and I should start looking for my red flares! **Fig 5.11**

Fig 5.11   Using the home-made rangefinder

Shooting up an Object

Very often you badly need the pin-point accuracy of a three bearing fix but you are only sure of two objects. There is a third, but due to lack of local knowledge you are unsure of its identity. The answer is to 'shoot it up'.

Plot a two-bearing fix using your two known objects. This will give you at least an approximation of your position. Now take a bearing of the unknown one. Plot this bearing on the chart outwards from your assumed position and it should pass through, or very close to, a feature on the chart resembling the one you wish to identify. **Fig 5.12** If it does you can rely on your fix.

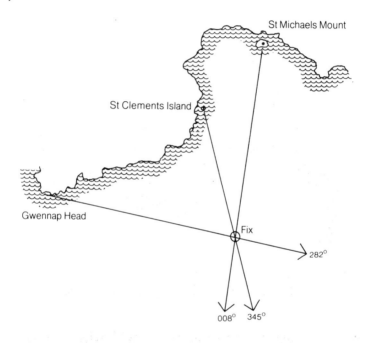

Fig 5.12   'Shooting-up' an object

Example. Gwennap Hd bears 282°, St Michael's Mt bears 008°. Unidentified small island bears 345°.

(a) Plot the fix using the two known position lines.
(b) Draw a bearing 345° from the fix.
(c) Name the island. Answer: St Clement's Isle.

And you can verify this position by using a range obtained by the methods described above.

Clearing an Object by a Given Distance

It is often necessary to work out a course which will clear an object by a given safe distance. For example, in **Fig 5.13** the vessel at A wishes to clear headland B by 2 miles.

   (a) With centre B draw an arc radius two miles.
   (b) From A draw tangent AC
   (c) AC is the course to steer.

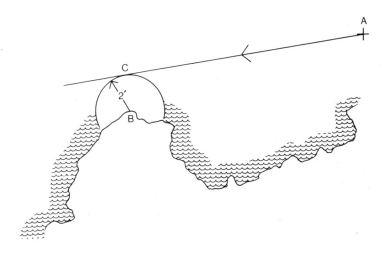

Fig 5.13   Clearing an object by a given distance

Now try plotting the chart run below. With what you have learned so far you should then be able to answer the six questions at the end.

CHART RUN ONE (Chart 5067)
1000   Trevose Hd Lt bore 105° range 3.5 miles. Fix.
        Course 224° Speed 10 knots.
1200   Godrevy Is Lt bore 163°.
        Godrevy Is Lt bore 105°. Fix.
1230   A/C (Alter course) to leave Pendeen Lt 2 miles on the port beam.
1315   Pendeen Lt bore 115° range 2 miles.

A/C to leave Longships Lt 1 mile on the port beam.

1350   Longships Lt bore 130°, Cape Cornwall bore 045°. Fix.

A/C 180°.

1400   Longships bore 048° range 1 mile. Fix.

A/C to leave the Runnelstone Buoy 5 cables on the port beam.

1420   Runnelstone Buoy bore 020°. A/C to leave Tater Du Lt 1 mile on the beam.

1440   Tater Du Lt bore 023°.

1455   Tater Du Lt bore 297°. Fix.

A/C for Low Lee Buoy (off Newlyn).

Questions

(1) What is course to steer at 1230?

(2) What is course to steer at 1315?

(3) What is course to steer at 1400?

(4) What is course to steer at 1420?

(5) What is course to steer at 1455?

(6) What is ETA (Estimated time of arrival) at Low Lee Buoy?

6 Lights and Buoys

One of the principal aids to pilotage is the system of lighthouses, lightvessels and navigational buoys around the coast. At night these are identified by the different characteristics of their lights which are described on the chart and it is important to understand them.

Many people prefer navigating at night, in the same way that many drivers prefer night driving. The reasons are that visibility is often better at night as there is usually less dust and water vapour in the atmosphere, the lights of land or those of another vessel are seen long before one would sight the same thing by day, and because of its characteristic one can instantly identify a lighthouse or lightvessel when during the day its identity might be less certain. Considerable experience is needed before entering strange harbours at night, but a night passage followed by an entry at first light has many advantages over the same passage made by day.

A quick glance at the chart reveals that each light has its characteristic described. In the case of Lizard Pt Lt (chart 5051) it reads:

Fl   3s   70m   29M

This is shorthand for a light that flashes every three seconds, 70 metres above MHWS, visible 29 miles on a dark, clear night.

Not only are you likely to see the Lizard Lt twenty-nine miles away, but under conditions of low cloud you may well see the 'loom' (the light reflected off cloud) at twice that distance, obviously an impossibility by day. Several times I have seen the loom of the Lizard at sixty miles and it is quite possible to cross from Devon to the Channel Isles accompanied in turn by Start Pt Lt, the Channel LV and Les Hanois on Guernsey.

You must always identify and verify every light, checking the legend on the chart with what you see. It is all too easy to assume that a lighthouse is the one for which we have been

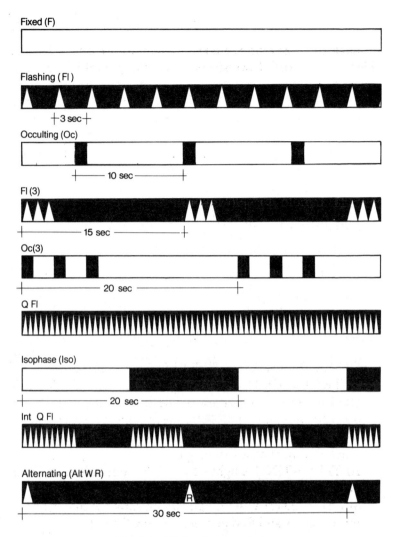

Fig 6.1 Light characteristics

looking but it can be a dangerous assumption.

You should at the passage planning stage write down the characteristic of each light you expect to see and the time it should show. If you are not on watch yourself you should give this list to someone who is and insist on being called when it shows and also if it fails to show at the expected time.

Back in the seventies a coaster went ashore under Trevose Head on the north Cornish coast. The skipper told the enquiry he thought he was near the Lizard! Now apart from anything else, Trevose shows a red light (Fl R 5 secs) and even in dense fog Trevose sounds a very different fog signal from the one on the Lizard.

The various types of light characteristic are illustrated in **Fig 6.1.** They are:

FIXED A continuous steady light.	Abbreviation F
FLASHING A regularly repeated flash.	Fl
OCCULTING A light which eclipses regularly.	Occ
ISOPHASE Equal light and darkness.	Iso

Other variations are:

GROUP FLASHING The light flashes in groups, in this case three.   Fl(3)
GROUP OCCULTING A light which eclipses in groups.
                                                                                        Oc(2)
QUICK FLASHING, VERY QUICK FLASHING and ULTRA QUICK FLASHING. Abbreviations, Q, VQ, UQ. Used mainly on buoys these define lights which flash 50–80, 80–160, and more than 160 times a minute respectively.
INTERRUPTED VERY QUICK FLASHING (IVQ) is the quick flashing version of groups.
ALTERNATING A light which shows one colour and then another, eg Al Fl W R 30 secs.
SECTORED (Fl W R) A light which shows White in one or more sectors and Red in other sectors. **Fig 6.2**

A sectored light is not, properly speaking, a type of characteristic. It is first and foremost a safety device. At any given point only one colour is observed. If you are in the safe sector then it shows white (or sometimes green), while if red shows it indicates that proceeding towards the light would take you over rocks or other hazards. Bearings of sectors, which are given in the Admiralty List of Lights are *always as seen by an observer from seaward.*

Should you use the alteration of colour as a position line?

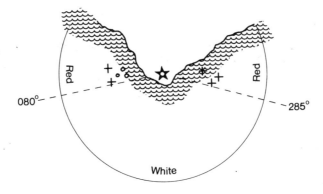

Fig 6.2 A sectored light

The answer is a very definite No! Fix only on the bearing of the light. The reason for this is that it is not often possible to detect the exact moment of colour change, and in any case atmospheric variations can cause a white light to adopt a reddish hue.

COLOURS We have already met with red, but other colours often used are:- Orange (O), Green (G), Blue (Bl), and Yellow (Y). There is also one other type of light whose characteristic is Mo (K). This is a morse flashing light, flashing K (— · —) at regular intervals.

EXERCISE FIFTEEN
Describe these lights in detail.

(a) F G
(b) Fl (3) 30 s
(c) Oc O 12 s
(d) Oc (3) R 12 s
(e) Iso R

(f) Q R
(g) VQ O
(h) Iso WR 10 s
(i) Al WR 30 s
(j) Oc WR 20 s

Buoyage

As with lights it is vital to identify a buoy. It is even more important because you are likely to be that much closer to a buoy and if you pass it on the wrong side you could be in danger. **Fig 6.3** Never make a fix on a buoy because buoys, unlike lighthouses, have been known to drift out of position, especially exposed ones such as the Runnelstone off Land's End. It is permissible to enter in the log 'Runnelstone abeam', but never use it as a basis for a position line, or a transit.

About twenty years ago one of the Navy's largest carriers grounded in Plymouth Sound although she was in the hands of a pilot. No damage was done, but she had to be dry-docked at enormous cost to the taxpayer. At the enquiry it transpired that a buoy was out of position and the navigating officer hadn't checked it. He was severely reprimanded. Incidentally, he would have checked it by 'shooting it up', as described in Chapter 5.

The buoyage of Northern Europe is now standardised as the International Association of Lighthouse Authorities (IALA) system and buoys consist of two main types – the Lateral and the Cardinal.

Lateral buoys are divided into two groups, port and starboard hand. Port are red and can shaped while starboard are green and cone shaped. They are to be left on your port or

It's OK lads. There's the West Wreck. (with apologies to Mike Peyton)

Fig 6.3   Always identify your buoy

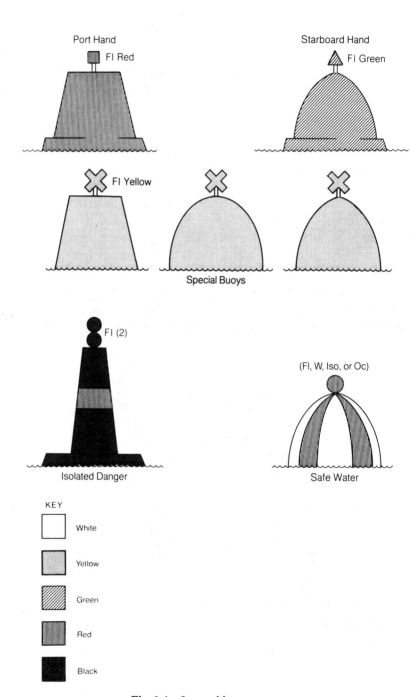

Port Hand — Fl Red

Starboard Hand — Fl Green

Fl Yellow

Special Buoys

Fl (2)

Isolated Danger

(Fl, W, Iso, or Oc)

Safe Water

KEY

White

Yellow

Green

Red

Black

Fig 6.4   Lateral buoyage

starboard hands respectively when entering a harbour, or travelling with the flood stream. Leaving a harbour or travelling with the ebb reverses the system. **Fig 6.4**

The flood stream is deemed to start at Land's End, dividing so that one half moves east up the English Channel and North Sea, and the other travels north towards Wales, the Scottish west coast, Pentland Firth and Moray Firth where it meets the southern flood. Here the buoys change over. For instance, as you journey north past Aberdeen you will have port hand buoys to port, but after leaving Thurso you will find starboard buoys on your port hand, just as you would voyaging south down the Irish Sea.

Cardinal buoys are all shaped the same, are black and yellow and have differing topmarks. They are used to mark a particular hazard, such as a rock or shoal. All cardinals flash a white light, whereas laterals flash red and green respectively.

Learning the lateral system is very simple. *Port hand* are *red cans* and show a *red light. Starboard hand* are *green cones* and show a *green light.* Once, all starboard hand buoys were black and I know of one that still is. It is in Plymouth Sound near another starboard hand buoy and is presumably left in its original livery in order to prevent confusion in fog.

Cardinal buoyage is nearly as simple. **Fig 6.5** All cardinals are pillar buoys and have largely replaced the old landfall buoys, such as the Runnelstone and Outer Spit. The way to learn them is by their topmarks. North cardinals have triangles pointing up, south pointing down. Westerly cardinals have two triangles with apexes touching. These form a Wineglass or WaspWaist shape (W for West). Easterlies are the opposite with two triangles with bases touching. (E for EGG-shaped.)

Now for the colour. As stated above, cardinals are always black and yellow and the topmarks point to the black. Thus north cardinals have black topsides and yellow lower parts, while south cardinals are the opposite. Westerly cardinals are black in the middle, with yellow tops and bases and easterlies are painted black top and bottom with yellow midriffs.

Finally, at night these buoys exhibit quick flashing lights, white. North cardinals show a continuous quick flashing light,

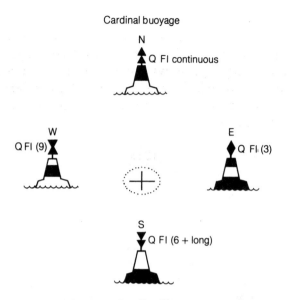

Fig 6.5   Cardinal buoyage

and then as you come *clockwise,* easterlies (positioned at three o'clock) show Q Fl (3). Southerlies (at 6 o'clock) flash six times, plus a longer flash. Westerlies (at 9 o'clock) flash nine times.

In addition there are special buoys which mark wrecks, spoil ground, (sewage outfalls, dumping grounds) or quarantine areas. They are always yellow and may be can-shaped, cone or spherical. Cans and cones should be treated as you would a similarly shaped lateral buoy, but spherical buoys may be left on either hand.

One to look out for is the Isolated Danger buoy which probably marks a solitary rock. It is red and black with two black circles for a topmark. Its light consists of two white flashes. At night special buoys exhibit yellow lights.

Safe water marks are white with red vertical stripes and show a white light, isophase or occulting.

A recent addition to the buoyage system is the preferred channel mark, really only intended for large deep-draught vessels. A red can with a broad green horizontal stripe indicates that the preferred channel lies to starboard so it should be left to port. It shows a red light (Fl 2 + 1) and is shown on a chart as a white can marked RGR. A green cone with a broad red stripe shows that the preferred channel lies to port and you should leave it to starboard. A green light (Fl 2 + 1 G) is shown at night and on the chart it is marked as a black cone GRG. **Fig 6.6**

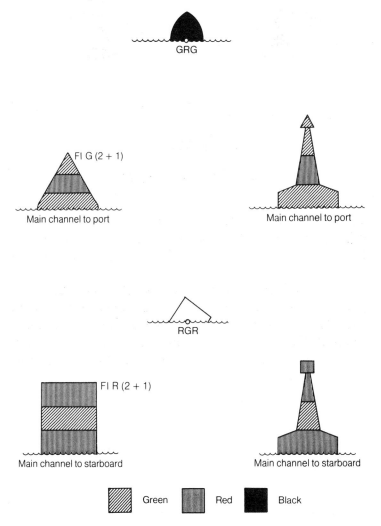

Fig 6.6   Preferred channel marks

EXERCISE SIXTEEN

(a) Describe a port hand buoy, indicating its light characteristic.

(b) Describe a starboard hand buoy indicating its light characteristic.

(c) If the buoy at (a) is sighted right ahead of you when entering a harbour what action should you take?

(d) Draw a cardinal buoy where the point of interest lies to the west, indicating its light characteristic.

(e) If you saw the buoy at (d) right ahead of you when you were steering 180° what action should you take?

7 The Rule of the Road

The conduct of all vessels navigating within sight of each other is governed by the International Regulations For Preventing Collisions At Sea. This embodies nearly forty regulations, each with many sub-sections which make pretty formidable reading. However, like so much else, the task can be simplified. First, it is important to remember that only about one third of the regulations lay down the action to be taken by approaching vessels. The remainder refer to the lights, shapes and sound signals to be shown or made by vessels so that the rules may be obeyed in all conditions of visibility.

To break down the first third further, there are only four basic principles which are:

The End-On Rule. Two power-driven vessels meeting end-on or nearly so shall both alter course to starboard. (The continental rule of the road.)

The Crossing Rule. When two power-driven vessels are approaching, the vessel that has the other on her starboard hand shall give way, like motor cars on roundabouts.

The Overtaking Rule. Any vessel overtaking another shall keep out of the way. Similarly the overtaken vessel shall maintain her course and speed (just as you would on the motorway). This rule, incidentally, applies to all vessels – a racing catamaran can easily overtake a VLCC. **Figs 7.1, 2 & 3**

The Not-Under-Command Rule. (NUC) Not under command does not mean the skipper is incapacitated. The vessel may be damaged, or suffering engine failure or may be a type of vessel which by its very nature is unable to get out of your way. Such vessels include sailing vessels, fishing vessels, and any which are restricted in their ability to manoeuvre, for instance a tug with a tow, a cable-layer, VLCC, or a warship undergoing replenishment at sea.

Fig 7.1 End-on rule

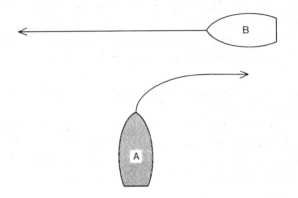

Fig 7.2 Crossing rule

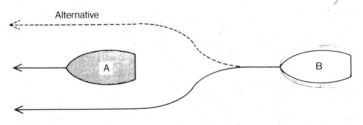

Fig 7.3 Overtaking rule

Giving way to another ship means either altering course, slackening speed, stopping or going astern. Think how many times you avoid a collision on the road merely by reducing speed. As on our crowded roads there often isn't room to alter course in a busy sea-lane or fairway. When crossing the Channel I rarely alter course to vessels approaching on my starboard hand. Very often it is simpler and less time is wasted to reduce speed until the other vessel is clear.

If you are going to alter course the rules state that you must do so in plenty of time and the alteration must be obvious. In other words an alteration of five degrees may take you clear of the other chap, but to him it will appear that you are merely yawing. At least forty-five degrees is recommended.

Why are privileges accorded to sailing vessels? Simply that they can never be as manoeuvrable as a powered vessel since they cannot sail into the wind, they can rarely make an emergency stop, and they cannot go astern. This is not to say that sailing vessels are not governed by their own rules when meeting each other. Three rules apply here also.

The sailing vessel which has the wind on her starboard side has the right of way.

When both vessels have the wind on the same side, the windward vessel shall keep out of the way.

A vessel running free (ie with the wind abaft the beam) shall keep out of the way of a close hauled vessel.

In addition, there is the overtaking rule referred to above.

A sailing vessel loses her priority to any vessel not under command, restricted in her ability to manoeuvre or constrained by her draught (as for instance a ship in a narrow channel) and to a vessel fishing.

Notwithstanding these rules it is also common sense and courtesy not to stand on your rights and force an alteration on any working vessel if you can avoid it. Also remember that the moment a sailing vessel switches on an engine, even if the sails are still hoisted, it becomes a power vessel under the regulations.

Collision Courses

Always take a bearing, usually with the hand-bearing compass, immediately you notice a vessel approaching on either hand. After a few minutes take another and then a third. If the bearing increases then you will pass ahead of her and if it decreases she will pass ahead of you. If the bearing remains constant, however, then a real risk of collision exists. **Figs 7.4 & 5** At the same time check your closing speed if you can on

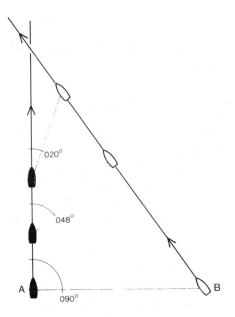

Fig 7.4 Collision courses

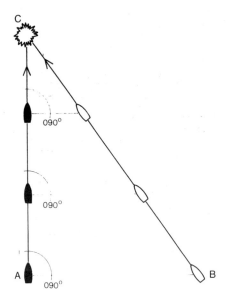

Fig 7.5 The result of a steady bearing

radar. A reduction of one mile in three minutes means a closing speed of 20 knots.

Relative Bearings

In Chapter 3 it was stated that compass bearings were always to be written as three digits, but bearings relative to the ship's head are a two figure estimate of angle. This is so that a lookout can make a quick report of an object to the skipper without either having to refer to the compass. These relative bearings are always prefaced by *red* or *green,* according to whether the object is to port or starboard and, provided the angle is less than 100°, is given as two digits (eg Red 10, Green 30). Red 90 for instance would be an object on the port beam. If the object is 'right ahead' or 'dead astern' that is the report made. It is also permissible, indeed recommended, to report 'On the port bow', 'Port beam', 'Starboard quarter', etc.

In the following diagram only starboard is shown. Port is identical but prefixed with red. **Fig 7.6**

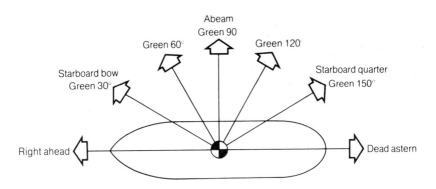

Fig 7.6   Relative bearings

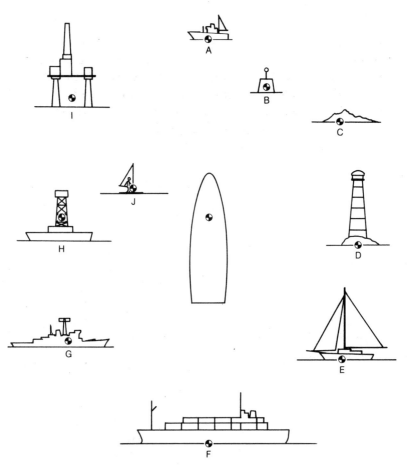

Fig 7.7   Relative bearing exercise

EXERCISE SEVENTEEN
Using the diagram in **Fig 7.7** give the relative bearings of each
object A–J. (Your vessel is in the centre.)

EXERCISE EIGHTEEN

In **Fig 7.8** your vessel is at the centre of the display. Imagine that you have sighted just ONE of the vessels A to I. What action do the rules instruct you to take in each case?

Clue: The answer will be either to alter course to starboard, or maintain your course and speed.

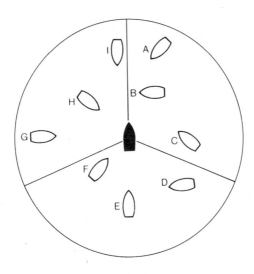

Fig 7.8   Exercise 18

8 Shapes, Lights and Sound Signals

During daylight hours certain shapes are shown by vessels to indicate to other ships what they are up to. They are almost always black and consist of four types: spheres, cones, diamonds and cylinders. They are hoisted in the forepart of the vessel where they can best be seen and the principal ones are:-

One black ball	Vessel at anchor.
Two black balls (vertically)	Vessel not under command.
Three black balls (vertically)	Vessel aground.
Two cones apexes touching	Vessel fishing (over 20 metres).
Basket	Vessel fishing (under 20 metres).
Black diamond	Tug with tow over 200 metres.
Cone apex down	Sailing vessel using engine.
Black diamond between black balls, arranged vertically	Any vessel unable to manoeuvre eg Cable-layer, Survey ship, Replenishing at sea, etc.
Black cylinder (at least 2 metres long)	Vessel constrained by her draught, eg VLCC (Very large crude carrier).

Fig 8.1

Having said that, it is worth pointing out that there are a few miscreants about who do not obey these rules to the letter. Notoriously, fishermen keep their cones and baskets permanently hoisted, even in harbour. Even though you may be absolutely sure they are not fishing, because they are flying their shapes you are duty bound to give way. The best that can be said for it is that they are working and you are probably not. But I have no time for the yachtsman who uses his auxiliary engine and fails to hoist a cone. His is, of course, a powered

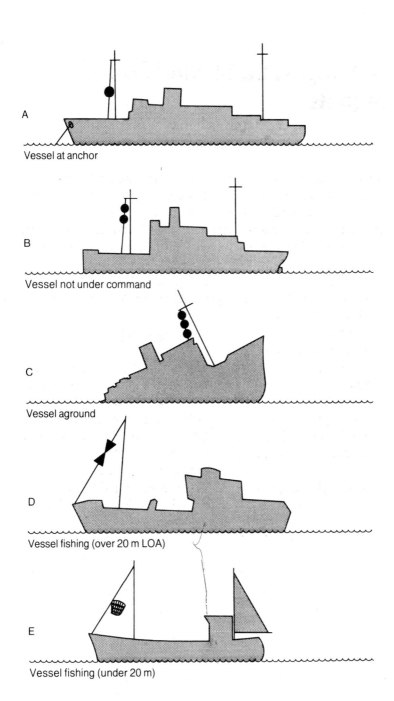

A Vessel at anchor

B Vessel not under command

C Vessel aground

D Vessel fishing (over 20 m LOA)

E Vessel fishing (under 20 m)

Fig 8.1 Shapes shown by vessels by day

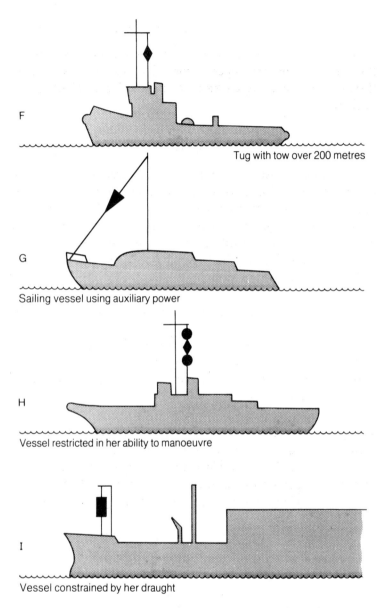

F

Tug with tow over 200 metres

G

Sailing vessel using auxiliary power

H

Vessel restricted in her ability to manoeuvre

I

Vessel constrained by her draught

Fig 8.1   Shapes shown by vessels by day

vessel and should be treated as such, but quite a number of them expect the privileges accorded to sailing vessels even though they are doing six knots and haven't a stitch of canvas in sight.

Lights

A ship's lights must be shown between sunset and sunrise and in bad visibility. Times of sunrise and sunset are given in the nautical almanac. Remember to allow for British Summer Time as all times given are GMT. As to what constitutes poor visibility, don't wait until you have a genuine pea-souper. A heavy rain squall is sufficient even if it only lasts a few minutes.

There are three types of light.

An all-round light, which is usually white but may be specified to be red or green.
A light sectored to show only over certain specified arcs. These are known as steaming lights.
Red and green sectored lights which are known as navigation lights.

The lights shown by a powered vessel under way at night are shown in **Fig 8.2** Under way, incidentally, means the vessel is not attached to the ground in any way and does not necessarily mean the vessel is moving. It could be under way but stopped. A moving vessel is always referred to as being 'under way and making way'. **Fig 8.3**

The arcs and ranges look complicated but in fact are not. All one has to do is remember the magic figure 225°. This is the arc for the steaming lights and if one halves that figure one arrives at 112½° which is the arc for the red and green navigation lights. Subtract 225° from 360° and you have 135°, the arc of the overtaking and towing lights.

As far as ranges are concerned, remember this is the minimum required in good visibility. Why not identical ranges for all lights? Well, if for instance all you could see of a large vessel's lights were her two steaming lights you could assume

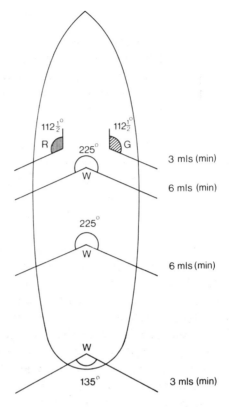

Fig 8.2   A ship's lights

you were between three and six miles from her. The moment you also made out the red or green navigation lights you would know you were within three miles. These ranges decrease with diminishing size of vessel. A vessel of less than fifty metres needs lights with ranges of 5 miles and 2 miles respectively and in vessels less than twelve metres all lights except navigation lights shall be 2 miles in range, the sidelights being 1 mile.

The easiest way through the maze of lighting regulations is to examine what is required by each type of vessel beginning with the smallest. **Figs 8.4 and 8.5**

Small rowing boats, dinghies, tenders are not required to show a permanent light, but if they do not they must have ready to hand a white light which must be shown in sufficient time to avoid a collision.

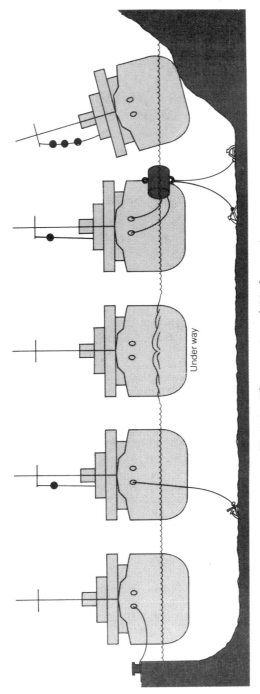

Under way

Fig 8.3 The meaning of 'Under way'

A yacht under 20 metres may show a combined lantern at the masthead, consisting of red and green navigation lights and a white stern light.

Yachts over 20 metres must carry separate navigation lights and a white sternlight. No masthead light is necessary. An option for large yachts is to carry additionally two all-round lights, red over green at the masthead.

Power-driven vessels must carry white lights at the masthead. Under 7 metres in length they may show an all-round white light above navigation lights. No sternlight is necessary in this case. Power-driven vessels under 50 metres must display a white steaming light, navigation lights, a stern light and a yellow towing light, this last having the same arc as the sternlight. Over 50 metres and they must display a second steaming light aft, higher than the forward one.

All vessels at anchor must display forward a white all-round light, visible at least two miles. A vessel over 50 metres will display a second all-round light aft, lower than the forward one.

A vessel not under command will show two red lights vertically, visible all-round. She will not show her steaming lights. If she is making way she will show her navigation lights and sternlight. When under way but stopped only her NUC lights will show.

A vessel aground over 20 metres will show her all-round anchor light(s) and red NUC lights, but nothing else.

A vessel unable to manoeuvre such as a cable-layer, will show in addition to her steaming and navigation lights, three vertical all-round lights, these being red, white and red. These correspond to the ball, diamond, ball of the daylight shapes. **Fig 8.6**

A VLCC or any vessel constrained by her draught will show, in addition to her steaming and navigation lights, three all-round red lights vertically.

A vessel trawling, in addition to steaming and navigation lights shall display two all-round lights, green over white.

Other fishing vessels (crabbers, long-liners) shall

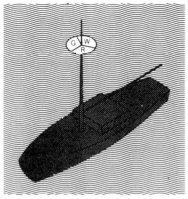

Yacht under 20 metres

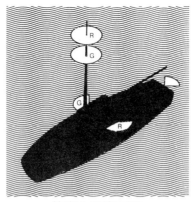

Yacht over 20 metres

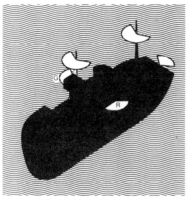

Vessel over 50 metres
under way and making way

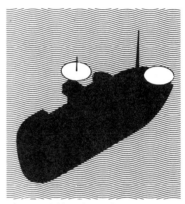

Vessel at anchor

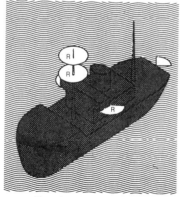

Vessel not under command

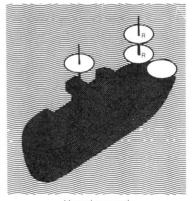

Vessel aground

Fig 8.4   Ships' lights

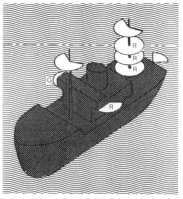

Vessel constrained by her draught

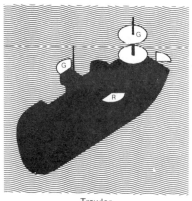

Trawler

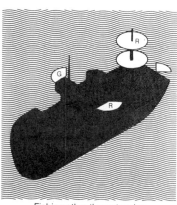

Fishing other than a trawler

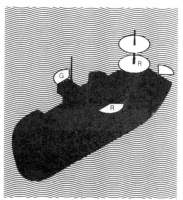

Pilot vessel

Vessel towing

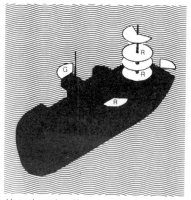

Vessel restricted in her ability to manoeuvre

Fig 8.5   Ships' lights

Air-cushion vessel

Fig 8.5 Ships' lights continued

additionally display two all-round lights, red over white. If any fishing vessel has gear extending more than 150 metres from the vessel she must exhibit a white light in the direction of the gear. There are additional rules in Annexe II of the Regulations referring to pair trawlers and purse seiners which will repay study.

A pilot vessel exhibits an all-round white light over red.

A vessel towing exhibits two steaming lights vertically and a yellow towing light above her sternlight. If the tow exceeds 200 metres she shows three white lights. The towed vessel only exhibits side or navigation lights and a sternlight.

A hovercraft will exhibit an all-round yellow flashing light in addition to normal steaming lights when in the non-displacement mode.

The object of all this is to enable you to answer the following questions day or night.

WHERE IS SHE? (Ahead, astern, port or starboard)
WHAT IS SHE? (Sailing, fishing, NUC)
WHAT IS SHE DOING? (At anchor, aground, towing)
WHAT IS SHE ABOUT TO DO? (Pass ahead, overtake, on collision course) and to take the appropriate action.

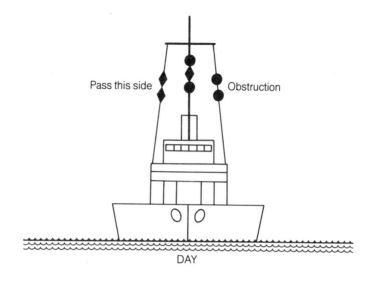

Pass this side   Obstruction

DAY

NIGHT

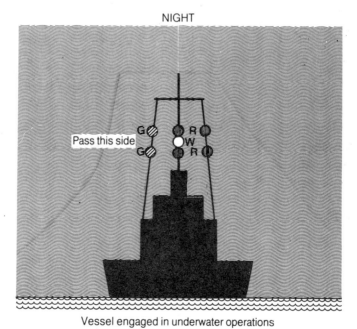

Pass this side

Vessel engaged in underwater operations

Fig 8.6   Vessel engaged in underwater operations by day and night

Sound Signals

The regulations require all vessels to carry some form of sound-making apparatus in order to warn other vessels of their approach, especially in poor visibility. In a small dinghy this might be nothing more than a referee's whistle, but most small craft will carry at least a compressed-air foghorn. It is worth remembering they are only good for about 200 to 250 blasts, but they are re-chargeable. Other types include air horns, working off the twelve volt battery, sirens, klaxons, bells and gongs.

Sound signals are divided into two types – those used in good visibility and those used in fog. The first group are only used in sight of other vessels, and there are six to remember.

ONE SHORT BLAST	I am altering course to starboard.
TWO SHORT BLASTS	I am altering course to port.
THREE SHORT BLASTS	I am operating astern propulsion. (This does not necessarily mean going astern.)
FIVE OR MORE SHORT BLASTS	Wake up! (the official definition is: I have doubt as to your intentions to avoid a collision, but the meaning is the same.)

The definition of a short blast is one of about 1 second's duration.

When overtaking in a narrow or busy fairway, a combination of long and short blasts is used. The duration of a long blast is from 4 to 6 seconds.

TWO LONGS plus ONE SHORT BLAST	'I intend to overtake on your starboard side.'
TWO LONGS plus TWO SHORT BLASTS	'I intend to overtake on your port side.'

The confirmatory reply to these signals, indicating that the vessel to be overtaken agrees, is ONE LONG, ONE SHORT, ONE LONG and ONE SHORT blasts.

Fog

In fog, however, it is a different story. There are still only seven signals to be memorised but as fog is the traveller's worst enemy, they are vitally important.

ONE LONG BLAST EVERY TWO MINUTES	A power-driven vessel under way and making way. ('Here I come, sounding one.')
TWO LONG BLASTS EVERY TWO MINUTES	A power-driven vessel under way but stopped. ('I have waited TOO long')
ONE LONG BLAST plus TWO SHORTS (D)	Any vessel not under command, fishing vessel, sailing vessel, towing vessel, vessel constrained by her draught, in other words any vessel which would find difficulty in manoeuvring.
BELL RUNG FOR 5 SECONDS EVERY MINUTE	Vessel under 100 metres at anchor. If the vessel is over 100 metres the bell is rapidly followed by the sounding of a gong in the after part. (This makes sense as a VLCC can be 400 metres long)
THREE STROKES ON A BELL FOLLOWED BY RAPID RINGING FOR 5 SECONDS FOLLOWED BY A FURTHER THREE STROKES.	A vessel aground.

FOUR SHORT BLASTS   A Pilot vessel (. . . . is Morse Code for H and by day pilot vessels fly flag H.)

EXERCISE NINETEEN

1 What do the shapes in **Fig 8.7** tell you about the vessels displaying them?

Fig 8.7   Exercise 19

2 What shape does a sailing vessel display to show she is under power?

3 A vessel exhibits three black balls in a vertical line. What does this tell you about her?

4 What do the following lights tell you about the vessels exhibiting them?
(a) Red over white all-round.   (b) Green over white all-round.   (c) Three red lights vertical.

5 What is the meaning of an all-round white light above an all-round red?

6 What type of vessel displays an all-round yellow flashing light?

7 What, in good visibility, do these sound signals mean?
(a) Three short blasts   (b) One short blast   (c) Two long blasts, plus one short.

8 In fog, you hear the following sound signals. What does each mean?
(a) One long blast every two minutes.   (b) One long blast plus two shorts.   (c) Rapid ringing of a bell for 5 seconds every minute.

9 In fog, you hear a bell ringing immediately followed by a gong. What does this tell you about the vessel sounding it?

10 In good visibility a vessel astern of you sounds 2 long blasts followed by 2 short blasts. What is she telling you, and what signal should you make if you are prepared to comply?

9 Distress Signals

It is sincerely hoped that you never have cause to make any of these signals, but others may, and it is criminal to go to sea unable to recognise them and be able to give assistance to others in distress, even if you think it will never happen to you.

A convenient way to list methods of indicating distress is in descending order of equipment. For example, one method which all students remember is the raising and lowering of arms, yet this is absolutely a last resort, to be used when all else has failed, in a liferaft or on an upturned dinghy.

So the first method to use, if available, is radio because it is going to be heard over a very wide area and should bring the quickest response. You will almost certainly use VHF Channel 16, and the preface is, as everyone knows from countless films, MAYDAY, from the French 'm'aidez' or 'Help me'.

It must be stressed that this must only be used in dire emergencies, such as imminent sinking or the abandonment of a vessel when human life is at extreme risk. For lesser emergencies, such as man overboard or engine failure, PAN is used. For a medical emergency, eg heart attack or fractured skull, the preface is PAN MEDICO.
The routine is this

MAYDAY   MAYDAY   MAYDAY.
This is . . . (Name of vessel three times)
MAYDAY . . . . . (Name of vessel once)
MY POSITION . . . . (Lat, Long or true bearing and range
FROM a known position)
. . . Nature of distress and type of assistance required.

Radio apart, the most effective way of indicating distress in good visibility is with pyrotechnics because they draw attention immediately. At night a red parachute flare is best, with a hand-held red flare second. By day, a red exploding flare will attract attention, as will a hand-held orange smoke canister. Schermuly market Inshore, Coastal and Offshore packs, so you can purchase one to suit your type of cruising.

In fog, when many accidents occur, sound signals must be used. Internationally recognised is VICTOR (. . . —), 'I require assistance', as is the continuous sounding of a siren.

In good visibility a large vessel might hoist the flags N C. Since this means 'No – Yes' in the International Code it is obviously meaningless and can only signify trouble. She might also hoist a black flag over a black ball, or anything resembling same, whilst flag VICTOR means the same thing.

Large vessels may also use their Long Wave transmitters which have a much greater range than VHF, and they will broadcast on 2,182 kHz. Similarly you could pick up SOS (. . . — — — . . .) which is instantly recognisable, and could be made from a small boat with a lamp or torch.

Burning oily rags in a barrel or bucket is recognised internationally as a distress signal, but only as a last resort as you probably have sufficient trouble already without starting fires.

Lastly, we come to the hand-waving. Don't wave back – go and pick him up!

You must also learn the signal for warning other vessels that they are heading for danger, as is your duty. This is the letter U by international code flag, or morse by lamp or sound. It could be flashed at *you* by another vessel, coastguard station or from a manned lighthouse.

In the 1960s the crew of the Seven Stones Light vessel flashed it at the *Torrey Canyon* before she grounded, and presumably it was made by passing vessels to the owner of a yacht which foundered on rocks near Land's End. She had sailed five thousand miles from Brazil in perfect safety, but the last twenty feet wrecked her. Her helmsman had gone below to make coffee and fallen asleep. Fortunately there was no loss of life.

Learn all the distress signals in **Fig 9.1**.

Fig 9.1   Distress signals

EXERCISE TWENTY
Which method of indicating distress would you use in the
following circumstances?

(a) By radio.
(b) At night, no radio.
(c) By day, no radio.
(d) In fog, no radio.
(e) On a raft or small open boat.

CHART RUN TWO Chart 5051
Fishing vessel *Pat* slipped from Newlyn and passed Penlee Pt
buoy (Low Lee) at 0900, steering 140° at 6 knots.
1130 A/C (Altered course) 080° and steamed for 4.5 miles.
She then altered course for a position 090° Lizard Pt, range 3
miles and streamed her nets.
At 1500 *Pat* hauled nets and steamed 070° at 6 knots for forty
minutes.
1530 Yellow buoy bore 029°, Black Hd bore 315°. Fix.
Pat then set course for Black Rock Bn (entrance to Falmouth.)

Questions
1 Give DR position (lat and long) at 1100.
2 Why did the skipper steer 140° and not 130°?
3 Give DR position after steaming 080° for 4.5 miles.
4 What course was steered to reach the position off Lizard Pt?
5 What was ETA (Expected Time of Arrival) at that position?
6 What course was steered at 1530?
7 What was ETA Black Rk Bn abeam?
Now attempt Chart Runs 5, 6 and 7 on p 168–9.

10 Allowance for Tides

The most important factor affecting the calculations of the navigator making coastal or offshore passages is the vertical and horizontal movement of the sea's surface which we call the tides. In land-locked seas like the Mediterranean this movement is negligible, Greek waters having a rise and fall of only half a metre, but our coasts are very different. The range at Newlyn averages over five metres at spring tides while the Channel Isles and Bristol average eleven and thirteen metres respectively. Since the water in question is going to ebb and flow over the same period wherever you are, namely about six hours, it follows that the tidal streams off St Helier will be faster than those off Corfu.

The gravitational attraction of the moon and to a lesser extent the sun is responsible for the tides. The moon exerts an attractive force of twice that of the sun due to its proximity. In **Fig 10.1** the moon is causing high water at A by its attraction, while high water at B is caused by centrifugal force due to the earth's rotation. At C there is low water as the earth rotates, so these highs and lows circulate round the surface of the earth.

Spring and Neap Tides

Spring has nothing to do with the seasons, but comes from the old Norse word 'sprungen', to leap, and anyone who has watched a spring tide streak across the flat sands near the mouth of the Humber will readily understand its origin. In the Bay of Fundy in Northern Canada where they have sixteen metre tides, the highest in the world, you could easily be overtaken trying to escape an incoming tide.

Spring tides occur in fortnightly cycles and are caused by the moon and sun pulling in unison, when the moon is new or full. Their opposite, neap tides, are caused by the sun and moon pulling at right angles as in **Fig 10.2**. This occurs when the

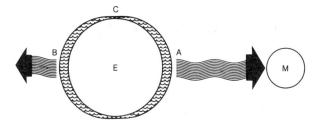

Fig 10.1   The cause of tides

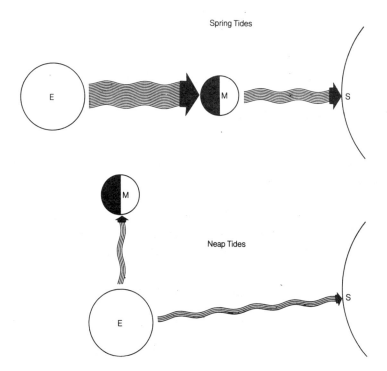

Fig 10.2   Spring and neap tides

moon is in its quarter and half phases.

A lunar day is about 24 hours and 50 minutes. A high tide occurs twice in that period, approximately every 12 hours 25 minutes. Thus the period between successive high and low waters is 6 hours 12 minutes.

The Twelfths Rule

The rate that a tide rises or falls is not uniform. It begins its ebb or flow slowly, accelerates to a maximum about half-way through its period and then slows again. A rough, but very workable, guide to the level of water at any time in a six-hour tide is known as the Rule of Twelfths, and it works like this.

Hour 1 $\frac{1}{12}$ of the range
Hour 2 $\frac{2}{12}$ of the range
Hour 3 $\frac{3}{12}$ of the range
Hour 4 $\frac{3}{12}$ of the range
Hour 5 $\frac{2}{12}$ of the range
Hour 6 $\frac{1}{12}$ of the range

For example, let us suppose that a harbour experiences a range of six metres. It follows that during the first hour a negligible rate of half a metre will occur. If you are secured alongside and are tending warps you will have little to exercise you. But two hours later a quarter of the entire rise, ie 1.5 metres, will occur and you will be kept busy. Even more important, if the tide is ebbing, those third and fourth hours are the ones when your lines will become bar-taut very rapidly.

It also follows that at sea and in estuaries the rate of flow of the tidal stream will be most rapid during that middle period. If you are under sail or have minimal power you would do well to avoid a foul tide at this time, especially at springs.

Tidal calculations are more accurately worked out using the Admiralty Tide Tables and similar information is in Reed's and Macmillan's Almanacs. You need to be very accurate if you intend to use the height of a lighthouse in order to find your distance off, bearing in mind that height is only relative to MHWS and not to the height of tide at any given moment. To the charted height you must add the amount by which the water level is below HWS. Similarly, the headroom of a bridge is relative to HWS and will normally be greater. In these cases the following equation must be used.

Clearance = (Height of bridge + Height of HWS) minus (Height of tide + Height of mast).

Example 1 A bridge is shown on the chart as being 9 metres above HWS. The distance from the waterline to your masthead is 11 metres. Will you pass safely under?

  (a) Find the height of the tide for the day and time in question. Say it is 1.9 metres above Chart Datum.
  (b) Find HWS for that place. Say it is 5 metres above CD.

Then clearance is

$(9 + 5) - (1.9 + 11) = 14 - 12.9 = 1.1$ metres clear.

Example 2 A bridge is shown as being 6 metres above MHWS. The height of your mast from the waterline is 8 metres. Height of tide is 4.2 metres and HWS is 5.0 metres. Find the clearance.

Clearance $= (6 + 5) - (4.2 + 8) = 11 - 12.2 = -1.2.$

You will foul the bridge by that amount. **Fig 10.3**

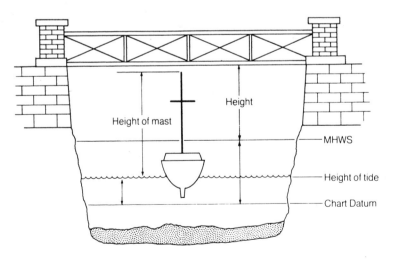

Fig 10.3   Clearance under a bridge

EIP-F

Chapter Six explained the expressions DR and 'Estimated' when applied to a position and to refresh your memory it is worth recalling that an estimated position is one which takes into account tidal stream and other factors. The example given was something like that in **Fig 10.4**.

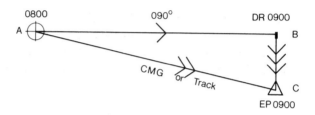

Fig 10.4   Effect of a tidal stream

The estimated position at C is all very well but it may be among rocks, shoal ground or even on the land itself, so we must make allowance for the tide BC and steer accordingly. To do this we construct a vector triangle, the three sides of which are made up of your intended track, the tide and finally a line representing your speed. When complete this third line is also your course to steer and the distance along the track where the two meet is your speed OVER THE GROUND or speed made good.

Example   Vessel at A, speed 6 knots wishes to reach B. There is a tidal stream setting 040°, rate 2 knots. What is the course to steer and speed made good?

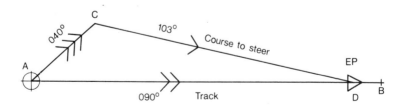

Fig 10.5   Allowing for the tidal stream

1 Join AB (which is 090°).

2 From A draw your tidal stream line 040° 2 miles long AC.

3 With centre C and radius your speed (6 knots) cut AB at D. CD is your course to steer and AD your speed made good. The course to steer is 103° and the speed made good is 7.5 knots. **Fig 10.5**

That was an example of a fair tide which added 1.5 knots to our speed over the ground. An example of a foul tide is in **Fig 10.6**. Once more we are at A wishing to get to B. Speed is still 6 knots but the tidal stream is now setting 250° at 1.5 knots.

Fig 10.6   Foul tide

1 Join AB (090° again).

2 From A draw tidal stream 250° 1.5 miles long. (AC)

3 With centre C and radius 6 miles cut AB at D.

CD is your course to steer and AD your speed made good. Course is 085° and we only make good 4.5 knots.

Add a time for A and a third piece of vital information becomes available, namely your ETA at B. Since D in both cases will be arrived at one hour after A, all that is left to do is measure DB and divide it by our speed made good. This will give us a period of time, which if added to time at D, say 0900, gives us an ETA at B.

In the first example our speed made good was 7.5 knots and DB was 1 mile. Time is Distance over Speed so $\frac{1}{7.5} \times 60$ (to bring it to minutes) = 8 minutes. ETA at B would be 0908. **Fig 10.7**

But in the second example speed is only 4.5 knots and we still have 3.8 miles to go. $\frac{3.8}{4.5} \times 60 = 51$ minutes so ETA is now 0951. Quite a difference! **Fig 10.8**

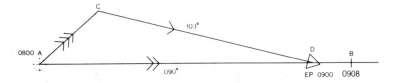

Fig 10.7   ETA with fair tidal stream

Fig 10.8   ETA with foul tidal stream

Tidal Definitions

The terminology used in **Fig 10.9** must be understood and familiarised if you are to make accurate calculations.

HEIGHT OF TIDE The height of the water level at any given moment above Chart Datum.

MEAN HIGH WATER SPRINGS (MHWS) The average of high water when spring tides occur. The level from which all charted heights, bridges and non-submersible rocks are measured.

MEAN LOW WATER SPRINGS (MLWS) The average of low water when spring tides occur.

SPRING RANGE The difference between the levels of the two averages above.

NEAP RANGE The difference between the levels of Mean Low Water Neaps (MLWN) and Mean High Water Neaps (MHWN).

CHART DATUM (LAT = Lowest Astronomical Tide) The lowest level to which tides can be predicted to fall in normal meteorological conditions. All Charted Depths (CD), drying heights of rocks and the Height of Tide are measured from this.

DEPTH OF WATER The height of Tide added to the charted depth.

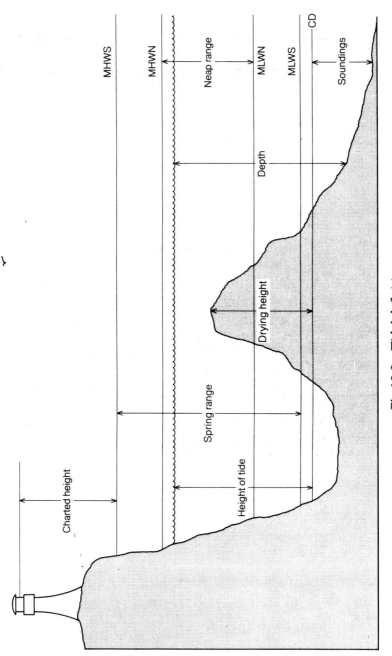

Fig 10.9 Tidal definitions

MHWS

MHWN

Neap range

MLWN

MLWS

CD

Soundings

Depth

Drying height

Spring range

Height of tide

Charted height

It is as well to remember that tide tables are a prediction. A long spell of high wind can cause the sea to heap up or push away from a particular coast. High, or low barometric pressure can lower or raise a tide by 0.3 of a metre. So you must always allow a margin of safety when calculating depths of water. The time may also vary from the prediction due to conditions.

EXERCISE TWENTY-ONE Chart 5051

1 0800 you are in position 50° 00′ N, 05° 00′ W (Point A).
   Set a course for 49° 50′ N, 04° 50′ W (Point B) allowing for a tide 090° rate 2 knots. Your speed 10 knots.
   What is (a) your course to steer, and (b) your ETA at B?

2 1200 you are in position 49° 50′ N, 05° 30′ W (A)
   Set a course for 49° 50′ N, 05° 10′ W (B) allowing for a tide setting 180° rate 2 knots. Your speed 10 knots.
   What is (a) your course to steer and (b) your ETA at B?

3 1800 you are in position 49° 50′ N, 05° 10′ W (A)
   Set a course for 50° 00′ N, 04° 50′ W allowing for a tide setting 270° rate 2 knots. Your speed 10 knots. What is your course to steer and ETA?

4 Which two phases of the moon do you associate with spring tides?

5 How long does it take for a succession of two high and two low tides to occur?

6 What is the datum used for
   (a) a drying height of 1.2 metres?
   (b) A depth of 5.2 metres?
   (c) A bridge 9 metres high?
   (d) A rock which dries 0.9 metres?

7 In **Fig 10.10** the labelled lines represent the following:
   any neap low tide, any neap high tide, Chart datum, MHWS and MLWS.
   Which line represents which level?

8 Calculate the clearance in the following circumstances.
   Height of bridge 11.0 metres.
   Height of tide 1.9 metres.
   Masthead height 7.5 metres.

Fig 10.10   Exercise 21 (7)

Height of MHWS 4.9 metres.
9   What clearance will be found in the following instance.
    Height of bridge 13.8 metres.
    Height of tide 1.0 metre.
    Masthead height 22.0 metres.
    Height of MHWS 5.5 metres.

Tidal Streams

How do we know what tidal streams to apply? In three ways which are

Tidal Atlas, which gives stream directions and strengths relative to HW Dover or some other principal port for various sea areas.
An almanac, which contains extracts from the above.
From the chart.

In various places on the chart you will notice small diamonds enclosing a letter. Turning to a corner of the chart you will find a table with columns labelled A, B, C, etc which gives directions and rates of tidal streams for that area for each hour over a thirteen hour period. Two rates are given, springs and neaps. If the calendar tells us we are in between, interpolation is necessary.

Running Fix – Allowing for the Tide

One last technique to learn before leaving the subject of tides and that is allowance for the tidal stream when using a running fix. You will recall that with a running fix its accuracy is dependent on a correct assessment of speed and the tidal stream must affect this.

Fortunately, with most outstanding headlands where one is forced to use a running fix since there are no other objects available, the stream will run parallel with the coast. Thus it will be either fair or foul and only speed is affected. But where one uses a rock lighthouse such as the Wolf Rock you may well be experiencing a tidal set at right angles to your course because the streams tend to be circulatory. If you examine chart 5051 about a mile to the north of the Wolf you will find the diamond enclosing the letter B. In the bottom right-hand corner of the chart in column B you will notice that in a thirteen hour period those streams run right around the compass, from 311°, through 088°, and 166° round to 286°. What is more their spring rate is often 2.0 knots or 2.2 knots as shown by the column beside the directions. So you can see how important it is to allow for tides when taking a running fix off the Wolf Rock.

Of course, one must keep a sense of proportion. The amount a tide will affect you is inversely proportional to your speed. In other words if you are only capable of six knots, then the tidal stream of two knots represents 30% of your speed, but a fast cruiser doing 18 knots is almost unaffected because the tide component is only 11% and offshore power-boat racers at 70 knots can afford to ignore it altogether.

Example   At 1100 Wolf Rk bears 330°. Our course is 260° and we are doing 8 knots. During the hour we are using the tidal stream is setting 165°, rate 2 knots. At 1200 Wolf Rk bears 045°. Fix ship at 1200.

1 Plot your 1100 position line, 330°.
2 Plot course of 260° and make it 8 miles long. This is your 1200 DR.

3 Add tidal stream of 165° 2 miles long. This is your EP at 1200.

4 Transfer your 1100 position line through your EP and plot your 1200 position line. Where it crosses your transferred position line is your Fix. **Fig 10.11**

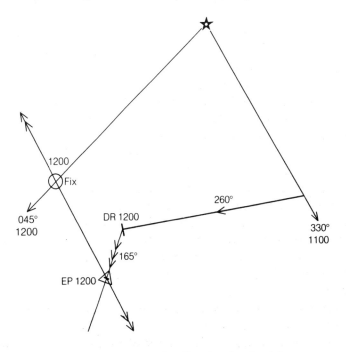

Fig 10.11 Running fix allowing for the tide

Echo Sounder and Lead Line

The instrument vital to your knowledge of the depth of water is the echo sounder. Once bulky affairs with elaborate graphs, they are now miniaturised with digital readouts, and some will sound an alarm if set to any chosen depth up to thirty metres. They come in a wide range of prices, but whichever type you use always remember that the transducer which sends its sound signal to the seabed is mounted in the hull, and you may have a metre or so of keel below that. The keel depth must be subtracted from the sounding to give depth of water beneath the keel.

Still widely used is the lead line. It is cheap, traditional, accurate and not subject to the vagaries of solid-state electronics. **Fig 10.12** shows how it is worked at intervals with pieces of bunting, thongs, etc to show different depths.

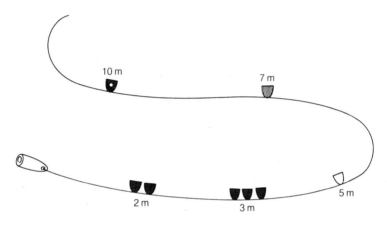

Fig 10.12   Lead line

In the base of the lead weight there is a shallow depression into which tallow may be pressed. This is known as 'arming the lead'. Its purpose is to collect samples of the bottom which can then be compared with that marked on the chart. If for instance it picks up blue clay when the chart shows the seabed under your supposed position as consisting of broken shingle then something is amiss.

Position Finding by Soundings

Once, in the Navy, I served under a captain who walked on to the bridge and ordered 'Navigate by echo sounder only for the next twenty-four hours'. And that's exactly what we did. Gyro-repeaters and direction finding equipment was switched off and no sights were taken. Only radar remained as an anti-collision precaution. We were well out of sight of land and shipping at the time and I should not like to attempt it in the English Channel, but it shows it can be done and in fog it could be a valuable way of navigating.

Problem You are approaching Helford River from the east. You are making good 270° at 6 knots when a fog bank looms up. Your 1400 DR is 50° 7.2′ N, 04° 55′ W. You decide to take soundings at fifteen minute intervals.

1 Take a strip of paper and fold it lengthwise.
2 Using the chart's scale mark on its edge intervals corresponding to the distance between soundings. In this case it will be 1.5 miles. (¼ × 6 knots).
3 Against each write the appropriate time and sounding.
4 Fold the strip over your parallel rule and set that to the course made good, comparing soundings with those on the chart. Don't forget to deduct height of tide from your soundings as those on the chart will be at datum.

It is highly unlikely that your figures will correspond exactly with those on the chart so you must slide your paper around, keeping to the same course made good, until there is a satisfactory relationship.

From the example in **Fig 10.13** we can see that we have been set into Falmouth Bay.

TIME	SOUNDING	TIME	SOUNDING
1400	60m	1445	47m
1415	61m	1500	10m
1430	60m	1515	17m

The position of the last sounding is your position.

A modern echo-sounder, whether digital or analogue, would identify the bank at 1500, dropping away below the 20 metre line at 1505, which would give a more certain identification of position.

I mentioned earlier that I wouldn't care to navigate by echo-sounder alone in the Channel, but in fact, when crossing to Cherbourg or the Channel Islands, I invariably note the DR

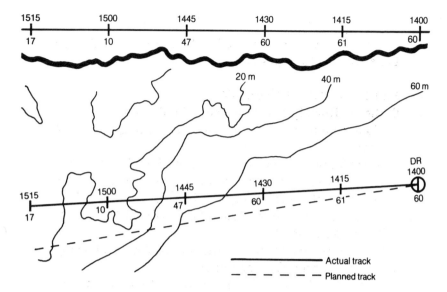

Fig 10.13   Navigation by soundings

time of the Hurd Deep (Chart 2675) and switch on the echo sounder about ten minutes before it is due. It provides one with a useful latitude, and sometimes, if the 'tongue' at 49° 25′ N, 03° 20′ W shows, a longitude as well. And if it fails to show I know I am on the ridge where the tidal diamond lettered J appears.

EXERCISE TWENTY-TWO Chart 5051.
(a) 0800 Lizard Pt Lt bore 040°. Your course 080°, speed 8 knots. Tidal stream setting 270° at 2 knots.
0900 Lizard Pt bore 290°. What is the lat and long of your 0900 fix?
(b) 1200 St Anthony Hd Lt bore 290°. Your course 190°, speed 6 knots. Tidal stream setting 040° at 2 knots.
1230 St Anthony Hd Lt bore 000°. What is lat and long of your 1230 fix?
(Remember to allow for only half an hour's run and tide in this and the following question.)
(c) 1500 Tater du Lt bore 250°. Your course 180° at 5 knots. Tidal stream setting 082° at 1 knot.

1530 Tater du Lt bore 282°. What depth is charted beneath your 1530 position?

CHART RUN THREE Chart 5067
2100 Bishop Rock Lt bore 000° range 3 miles. Fix.
      Your course 080° at 6 knots. Tide setting 030° at 2 knots.
2300 Seven Stones LV bore 013°, Round Island Lt 298°. Fix.
      Set a course to clear the Runnelstone buoy by 1 mile allowing for a tide setting 030° 2 knots.
0130 Longships Light bore 330°, Wolf Rock Lt bore 260°, Tater du Lt bore 028°. Fix.
      Set a course to clear the Lizard Pt Lt by 2 miles allowing for a tide setting 210° at 2 knots.
0430 Lizard Pt Lt bore 020°. Your course 080° at 6 knots. Tide slack.
0500 Lizard Pt bore 314°. Fix.

Questions
1 What is your EP at 2300?
2 What is your course to steer at 2300?
3 What is ETA Runnelstone abeam?
4 What is shown on the chart nearest your 0130 Fix?
5 What is your course to steer at 0130?
6 What is ETA Lizard Pt abeam?
7 What is the lat and long of your 0500 Fix?
You may now if you wish attempt chart runs 8, 9 and 10 in the last chapter.

11 The Compass

The compass is the most important piece of navigational equipment you have in your boat. So far all our thinking, and all our exercises, have been on the assumption that the compass pointed to True North as it does in big ships fitted with gyro compasses. It is much more likely that you will be using a magnetic compass, the needle of which points to the Magnetic North Pole which is somewhere in Northern Canada.

It is important to remember there are three 'Norths'. First, there is True North, which is the direction in which the needle of a gyro compass points.

Second, there is Magnetic North which is where the needle of a hand-bearing compass points in a magnetically simple boat held well away from any ferrous metal, gasbottle, radio, beer can or even a seaman's knife.

The difference between the two is called VARIATION. This changes with time and at present in West Cornwall for instance it is about eight degrees West, and decreasing by about five minutes annually. Variation is always shown on the compass rose of a chart.

The third 'North' is the one to which an uncorrected magnetic compass points and it could be anywhere. You have only to place some ferrous metal, a radio, or an electric motor near a small pocket compass to see the effect. The difference between this and magnetic North is DEVIATION. The two combined, ie allowing for both Variation and Deviation is called COMPASS ERROR as **Fig 11.1** illustrates.

The cause of Variation is the earth's magnetic field. If you can imagine an extraordinarily powerful bar magnet near the earth's centre with its blue pole pointing North, but a little out of alignment with the earth's axis then you can visualise lines of force emerging from somewhere in the Antarctic to pass over the earth's surface to return to the blue pole via Hudson's

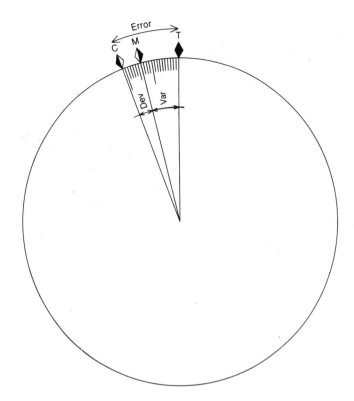

Fig 11.1   Compass error

Bay. Now suspend needles in this field and it is clear that whereas near the equator they would lie parallel to the earth's surface, in the vicinity of the poles they would appear vertical. In intermediate latitudes, such as 50° N they would tilt downwards towards the magnetic pole. Hence the difference in variation in different parts of the world. **Fig 11.2**

Even over such a comparatively short distance as the English Channel it can be seen from the chart that there is a difference of nearly three degrees between Cornwall and Beachy Head. On short passages one usually keeps to one figure for variation, generally a mean between the two. For example, on passage from Plymouth to the Isles of Scilly it would be safe to use 8 degrees throughout and on a crossing to Cherbourg, when Plymouth is 8° and Cherbourg is 6° a mean of 7 degrees

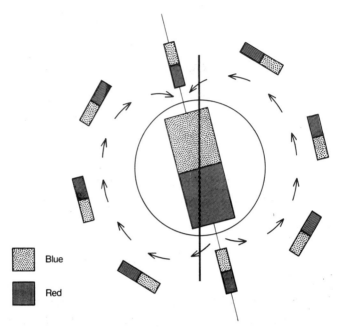

Fig 11.2 Cause of variation

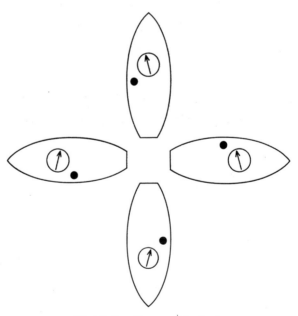

Fig 11.3 Cause of deviation

west for the whole trip could be used.

Deviation is a little more complex because it changes with each new heading. Its cause is a little more complex, too. In a GRP craft with no engine, no radio, no ferrous fittings whatsoever, there should be no deviation. But we have engines, generators, radios, fire extinguishers, standing rigging, and we carry knives and leave beer cans around the cockpit. All these affect a boat's compass to a greater or lesser degree, and as **Fig 11.3** illustrates, deviation will alter from east to west, and in magnitude, according to which way the ship is heading.

What can we do about it? Two things. You can go to a professional compass adjuster who, using magnets and soft iron masses, will reduce the deviations to manageable proportions and provide you with a deviation card for twelve points of your compass. His fee is not great, but the job has to be repeated after every lay-up.

Alternatively, you can do it yourself. A rough and ready method is to compare the steering compass with the hand-bearing compass while the latter is held well away from objects likely to cause deviation. By calling the ship's head on eight points of the compass and noting the differences, an approximation to a deviation card may be made out. But the correct method is to 'swing the ship' and this is how it is done.

First, the following precautions should be taken.

Pick a calm day with little or no swell. If possible aim for slack tide.
Use a large scale chart.
Ensure the vessel is in full trim so that no ferrous metal objects or radios will be added after swinging.
Ensure all radios, batteries or other electronic gear are at least two metres from the compass.

Now take your vessel out into your chosen spot and find a suitable buoy or beacon. Once there, identify an object ashore that is about five miles from your position, and is also on your chart. Plot its magnetic bearing from your position. Next, keeping as close to the mark as possible within the bounds of safety, steady the craft on a heading of 000° and read off the bearing of the fixed object. The difference between your reading and the charted magnetic bearing will be your deviation ON THAT HEADING.

Now steady up on 045° and repeat the procedure. Have an assistant write down your readings and continue for the remaining six points of the compass. If you have prepared a proforma before you start, you should end up with something like this.

Germoe
At low lee buoy Mag bearing Church Tower 085° Date..

Boat's Head	Bearing	Deviation
000°	078°	7° E
045°	083°	2° E
090°	086°	1° W
135°	091°	6° W
180°	092°	7° W
225°	090°	5° W
270°	084°	1° E
315°	080°	5° E

You will notice that when the compass read low the deviation was easterly and when high, westerly. This illustrates the principle behind all applications of deviation and variation and provides a mnemonic.

ERROR WEST, COMPASS BEST (ie high)
ERROR EAST, COMPASS LEAST.

The above information can be transferred to graph paper as in **Fig 11.4**. Such a graph is not very convenient to refer to quickly onboard so your next step is to construct a Deviation Card which can be pinned above your chart table. You do this by extracting as many points of the compass from the graph as you feel necessary and the result should look rather like this.

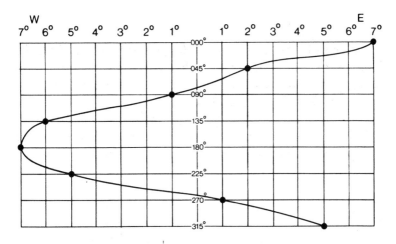

Fig 11.4 Deviation graph

Deviation Card   Yacht......................Date........................

Yacht's head compass	Deviation	Yacht's head magnetic
000°C	7° E	007°M
020°C	4° E	024°M
040°C	3° E	043°M
060°C	1° E	061°M
080°C	½° W	079½°M
100°C	2½° W	097½°M

and so on.

EXERCISE TWENTY-THREE
(a) What is the cause of Variation?
(b) What is the cause of Deviation?
(c) What is meant by compass error?
(d) In **Fig 11.5** say which of the angles is the Deviation. Is it Easterly or Westerly?
(e) Now do the same for **Fig 11.6**.
(f) Describe the construction of a deviation card.
(g) What do the letters M and C after a course or bearing indicate?

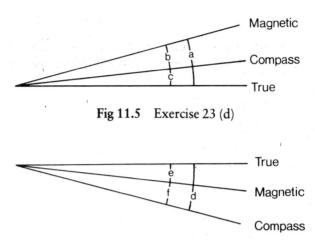

Fig 11.5   Exercise 23 (d)

Fig 11.6   Exercise 23 (e)

(h) Name the two errors you must correct to convert compass to True.

I described the method of swinging the ship when there was a useful mark such as a buoy. If there is no such thing handy the answer is to find a transit as far away as possible and steady the craft up on various compass headings. In Torbay, for instance, a suitable one might be the left-hand edge of Ore Stone in transit with the right hand edge of Thatcher Rock.

We must now learn to apply these errors so that all lines on the chart are relative to TRUE North. Always *draw* True courses, but *label* them with COMPASS courses to steer in order to avoid confusion by the helmsman. Similarly always convert a compass bearing to TRUE before laying it on the chart. This is in order to show your True position.

If both Variation and Deviation are Westerly then simply add the two together and add the total to a True course to give a Compass course to steer. (Error West, compass best.) Similarly, if the errors are both Easterly, subtract the total from the True course (Error East, compass least).

Example   Variation 8° W, Deviation 2° W, Total error 10° W. Add 10 degrees to the true course to give a compass course to steer. But if variation was 8° E and deviation 2° E then the total error is still 10° but it is now Easterly. Since 'Error East, compass least' subtract it from True to give a Compass course.

EXERCISE TWENTY-FOUR
(a) True Course 240°, Variation 8° W, Deviation 3° W. What is Compass course?
(b) True Course 240°, Variation 8° E, Deviation 2° E. What is compass course?

So far, both components have been in the same direction. What happens if one is east and the other west? Simply subtract the smaller from the greater and the name of the error is the name of the greater, thus:-

Variation 8° W, Deviation 2° E, Result is 6° and the error is W = 6° W
Variation 6° E, Deviation 3° W, Result is 3° and it's easterly = 3° E

EXERCISE TWENTY-FIVE
(a) True Course 110°, Variation 7° W, Deviation 4° E. What is Compass course?
(b) True Course 300°, Variation 4° E, Deviation 4° W. What is Compass course?

EXERCISE TWENTY-SIX
Now try these table exercises by filling in the missing gaps. The first one is done for you.

	True course	Vari-ation	Magnetic course	Deviation	Compass course
	180°	9°W	189°	2°W	191°
(a)	210°	?	200°	3°E	?
(b)	?	10°E	?	2°E	070°
(c)	091°	6°W	?	2°E	?
(d)	?	1½°E	?	9°W	077½°

When taking a compass bearing you must convert it to true before plotting it. Merely reverse the procedure you have been following, still remembering the mnemonic Error West, Compass Best, Error East, Compass Least.

EXERCISE TWENTY-SEVEN
Now complete this table. Again the first one is complete.

	Compass bearing	Devi-ation	Magnetic bearing	Variation	True bearing
	090°	3°W	087°	7°W	080°
(a)	200°	2°E	?	6°W	?
(b)	115°	?	117°	?	108°
(c)	?	½°E	348½°	11°E	?
(d)	195°	?	199°	18½°E	?

EXERCISE TWENTY-EIGHT
(a) Lizard Pt Lt bore 020° C, Variation is 8° W and Deviation 3° E. What is TRUE bearing?
(b) Longships Lt bore 165° C, Variation is 13° W and Deviation 7° E. What is TRUE bearing?
(c) Wolf Rk Lt bore 270° C, Variation is 12° E and Deviation 2° W. What is TRUE bearing?

Using chart 5067 work these examples

(d) Wolf Rock Lt bore 000° C, range 2 miles. What is your compass course to clear Lizard Pt by 2 miles? Variation is 8° W and Deviation on this heading is 2° E.
(e) Longships Lt bore 134° C, range 3 miles. What is your compass course to steer for Peninnis Pt (Isles of Scilly)? Variation is 8° W and Deviation on this course is 3° W.
(f) Lizard Pt Lt bore 083° C, Mullion Island bore 026° C. What compass course must you steer for Penzance harbour? Variation is 8° E and Deviation on that heading is 1° W.

CHART RUN FOUR Chart 5051

Deviation Card		Variation 8°W
Compass	Deviation	Magnetic
000°C	3°E	003°M
020°	4°E	024°M
040°	5°E	045°M
060°	5°E	065°M
080°	5°E	085°M
100°	5°E	105°M
120°	4°E	124°M
140°	3°E	143°M
160°	nil	160°M
180°	2°W	178°M

Only half the card is given as in this exercise all bearings and courses are in this sector. Remember that while the course remains steady the deviation will not change.

At 0800 Low Lee Buoy bore 270° C, range 1 mile. Your course 140° C. Speed 5 knots. 0830 St Michael's Mt bore 342° C.
Cudden Pt 020° C.
Porthleven Lt 081° C. FIX
A/C to clear Predannack Hd by 3 miles.
1030 Predannack Hd bore 095° C, Porthleven Lt 031° C, Lizard Hd Lt 115° C. FIX
1130 Lizard Pt Lt bore 055° C, A/C 092° C, Tidal stream 260° T, rate 1 knot.
1230 Lizard Pt Lt bore 314° C. FIX

Questions

1 What was your lat and long at 0830?
2 What was your compass course to steer at 0830?
3 What was your lat and long at 1030?
4 What was your lat and long at 1230?

Now attempt Chart runs 11, 12 and 13 on pp 171–3.

12 Communications

Communications at sea can be made by radiotelephone, radio broadcasts, signalling by lamp or sound signals, and flags. It is vital that all voyagers know the rudiments of each as the safety of their vessels, crews and the safety of others depends on it.

Communications fall into two groups, active and passive. Active signals are those initiated by you, such as 'Mayday', or more hopefully, permission to enter port. Passive signals are those received by you such as gale warnings.

Even if your boat has no radio equipment you should take to sea a portable set capable of receiving Shipping Forecasts on 200 kHz (1,500m). Beside it keep a timetable of weather forecast broadcasts and a proforma listing the sea areas and reporting stations so you can make a rapid note beside them. Some navigators keep miniature maps of all sea areas and note conditions in each, but I am not sure that this is essential, unless one wishes to be one's own meteorologist and attempt a forecast from the data provided.

SHIPPING FORECASTS 200 kHz
0555   1355   1750   0033

It is not a bad idea to set an alarm clock to go off five minutes before each forecast, even during the day when you are all around on deck.

When filled in the proforma might look something like this.

Date/ Time	Wind Dirn	Force	Weather	Visi- bility	Pressure
Portland	W	7	Sh	Mod	
Plymouth	W	6/8	Rain	Poor	
Lundy	WNW	8	Sh	Mod	
Channel LV	W	7	Rain	5 mls	996↘
Jersey	NW	6	Sh	12 mls	998↗

So a broadcast which described conditions on Scilly as being Westerly, force 6 veering NW and increasing to 8, Hail, 8 miles, 1006, rising, I would translate as SCILLY WNW 6/8 Hail 8 miles 1006 and remain in harbour.

A gale warning is issued when mean wind speeds of Force 8 (34 knots) or more are expected. The sea area will be referred to, followed by the words 'imminent' or 'soon' or 'later'. These mean within 6 hours, 6 to 12 hours, and more than 12 hours respectively. While we are on this unpleasant subject it is perhaps worth noting that Force 9 or severe gale refers to wind speeds of 41 knots while Force 10 is a Storm which implies gusts of at least 48 knots.

The forecasts also give information about visibility using the following definitions.

Moderate visibility	3.5 km or less. (2 miles)
Poor visibility	2.0 km or less. (1 mile)
Fog	200 metres or less. (1 cable)
Thick fog	50 metres or less. (50 yards)
Dense fog	25 metres or less. (25 yards)

Inshore Waters Forecasts

Immediately after the close of the 0033 Shipping Forecast Radio 4 transmits information for small craft operating in inshore waters. This is often more valuable than the area forecast as it is detailed and refers to local conditions.

Radiotelephones

The need for small craft to be able to communicate with each other, the coastguard and harbour authorities over a small radius (say 25 miles) gave rise to the invention of the VHF Radiotelephone. It is small, simple to operate and has little current drain. The following Channels are mandatory for its use.

CHANNEL 16   Should always be open when at sea. It is the distress frequency and all other vessels and shore stations should be listening out on it too. You may call up another vessel on it, but must go immediately to an agreed working Channel, clearing 16. Whatever you do, avoid chatting on it. Some fishermen with foreign accents are notoriously bad at this, and invariably collect 'une puce dans leurs oreilles' from the coastal stations for their pains.

The procedure is this.

VEGA (on 16) 'Sirius, this is Vega. Over.'
SIRIUS (also on 16) 'Vega, Channel 6, Over.'
VEGA (still on 16) 'Sirius, 6, Over'.

Vega will then switch to Channel 6 and call up again on that, releasing Channel 16 for further traffic or a distress call. If you are told to switch to a channel you do not possess, say so and list those that you do. Until recently it was also necessary to call Coastal Stations on Channel 16 and revert to a working channel, but this has now changed, and you must look in the Almanac to find the channel(s) each station requires you to use. For example, Niton uses Channels 4, 28 and 81, Jersey 25 and 82 and Land's End 27, 64, 88 and 85.

However, you still make initial contact with the Coastguard on 16.

CHANNEL 6   Mandatory. The primary inter-ship frequency.

CHANNELS 12 & 14   These two channels are normally used when calling port authorities.

CHANNEL 37   Most marinas can be called and will answer on this channel.

CHANNEL 67   The Coastguard's Channel.

PROCEDURES

The newcomer, listening to the expert, with his laconic and functional use of language, is sometimes hesitant about breaking into this rather exclusive club of professionals. But the growth of CB radios has accustomed more people to this new language, and it takes little learning but plenty of practice.

Its essence is brevity. Note in the example above just how few words were actually used. In fact, since the Falklands conflict the Royal Navy has refined and compressed its procedures in the light of experience, to the absolute minimum. For example, before 1982 if a ship wanted a radio check on volume and clarity the exchange of messages would be:

'Flag, this is Oscar Yankee. Radio Check. Over'.
'Oscar Yankee, this is Flag. Loud and clear. Out.'

Today the same exchange would be:

'Flag, Oscar Yankee. Radio Check. Over.'
'Oscar Yankee. Roger. Out.'

The fact that Flag has said 'Roger' means he has heard you loudly and clearly and it's all been done in 60% of the time.

PHONETIC ALPHABET
The phonetic alphabet was devised to avoid confusion between letters in messages which could be vital. In World War I this confusion cost lives so the first phonetic alphabet was invented and has undergone several revisions since, latterly to make it compatible between services of the NATO countries.

It is extremely easy to learn, especially if one looks for some logic or reason behind the choice of words. I have found that the easiest way to learn them is to group them whenever possible, thus:

ALFA   ROMEO (like the car)
ROMEO   JULIET (like the play)
BRAVO DELTA ECHO KILO PAPA SIERRA VICTOR YANKEE (Yanqui) (the Latin ones)
CHARLIE MIKE OSCAR (possibly international performers Chaplin, Hawthorn, and Peterson)
FOXTROT TANGO (Dances)
GOLF HOTEL WHISKY (Suggesting relaxation)
INDIA LIMA NOVEMBER QUEBEC ZULU (Geographical)

UNIFORM X-RAY (They don't fit anywhere else)

The important thing to remember is that the alphabet should be used as sparingly as possible. If the recipient of your signal can understand it perfectly in plain language there is no necessity to resort to phonetics. It should only be used, either when the word is so exotic that you assume the recipient will need a spelling, or if he requests one, as in the following imaginary exchange.

> 'Brixham Coastguard, Oscar Yankee. I have a motor vessel PISTOLE on fire. Position 200 degrees, Bolt Hd, 4 miles. Over'.
> 'Oscar Yankee, say again word after vessel. Over.'
> 'Brixham, I say again word after vessel, PISTOLE. I spell. PAPA INDIA SIERRA TANGO OSCAR LIMA ECHO. PISTOLE. Over'.

The other convention is to preface any numbers used (as in 200 degrees) with the word 'figures'. When using fractions, refrain from using 'point'. It is better to say 'decimal'.

The last three vital pieces of radio procedure are

MAYDAY	Only used when life is in imminent danger.
PAN	An emergency not yet of Mayday proportions. Man overboard, fire.
PAN MEDICO	A medical emergency.
SECURITAY	A navigational warning normally issued by a coastal station. Refers to changing weather conditions, floating wreckage, etc.

EXERCISE TWENTY-NINE
(a) On what BBC frequency are the shipping forecasts?
(b) Give the times of shipping forecasts.
(c) What information is given in reports from coastal stations that is omitted from forecasts?
(d) If a gale is forecast as being 'soon', what period of time is referred to?
(e) What is the name for a wind of force 10, and what is its wind speed?

(f) What is the definition of 'Poor' visibility?
(g) Where and when would you tune in for Inshore waters forecasts?
(h) Give the numbers of: the Distress Channel, the Port Authorities Channels, and the Coastguard's channel.
(i) Spell MORLAIX using the phonetic alphabet.
(j) What are the meanings of: Pan, Mayday, Securitay?

The International Code

Few small vessels carry a complete set of international code flags and you are unlikely to use more than about nine of them, although you should be familiar with about a dozen. This is because they affect your safety and those of other vessels. **Fig 12.1**

All these single letter signals can also be made by lamp or sound in Morse. Those you must be able to recognise are:

ALFA — I have a diver down. Keep clear of me at slow speed. (Failure to recognise this could result in the man's eardrums being damaged or worse. This is now more usually a painted board, as in calm weather the flag is apt not to be seen.)

BRAVO — I am carrying, loading or discharging dangerous cargoes. (Tankers, and vessels carrying explosives and chemicals fly this. Don't light up.)

DELTA — Keep clear of me. I am manoeuvring with difficulty. Flown by survey ships, cable layers or any vessel not under command.

HOTEL — Flown by Pilot vessels or any vessel carrying one.

OSCAR — Man Overboard. (Help him search.)

QUEBEC — My vessel is healthy and I request free pratique. (Flown until cleared by HM Customs.)

TANGO — Keep clear of me. I am engaged in pair trawling.

UNIFORM — You are standing into danger. (Vital you can recognise this.)

VICTOR — I require assistance. (Another vital one as it is one way of indicating distress.)

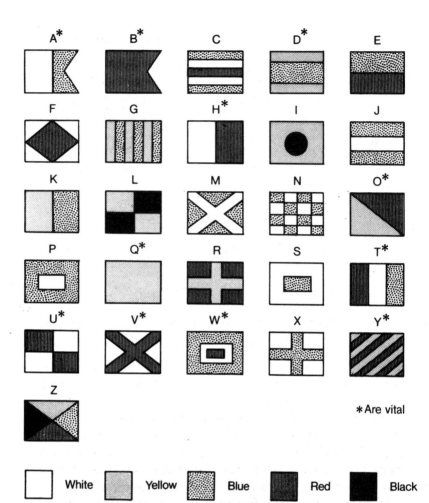

Fig 12.1   The international code of signals

WHISKY   I require medical assistance.

YANKEE   I am dragging my anchor. (And if you're in the same anchorage as he is you might be better off riding it out somewhere else.)

The flags which you may need to have aboard are:

ALFA   If you are doing any sub-aqua work or play.

DELTA   In case you should become disabled in a fairway.

GOLF   It means 'I require a Pilot', and even if you have radioed for one will help him find you.

OSCAR   For emergency.

QUEBEC   My vessel is healthy and I require free pratique. (A legal requirement to fly this when you return from abroad until cleared by customs.)

VICTOR   I require assistance.

WHISKY   For emergency. In case of accident or serious illness on passage.

I have not included YANKEE, because, since you wish to alert everyone to the fact that your anchor is dragging you are much better off making the sound signal YANKEE (— . — —). Many years ago I was woken at three in the morning on a very rough night by that signal made by a merchantman sheltering in Mount's Bay. Several vessels had sheltered there from a westerly gale, which had backed southerly, and one of them was warning her companions that her anchor was no longer holding. I doubt if her flag would have awakened me, nor attracted the attention of the anchor watches of the other vessels.

Those left out of the list of flags are as follows:

CHARLIE   Affirmative. (This would only be made in reply to a signal.)

ECHO   I am altering course to starboard. (Normally made by sound.)

FOXTROT   I am disabled. Communicate with me. (In that condition, he would more likely to be making one of the distress signals.)

INDIA   I am altering course to port. (As for Echo.)

JULIET   I am on fire. Keep well clear. (Probably only flown by a large vessel whom you couldn't help anyway.)

KILO   I wish to communicate with you. (Likely only to be flown by a Navy vessel who couldn't raise you on VHF or by lamp.)

LIMA   You should stop your vessel instantly. (As for Kilo.)

MIKE   My vessel is stopped and I am making no way through the water. (Of minor interest.)

NOVEMBER Negative. (As for Charlie.)

PAPA   Vessel about to proceed. (At sea) My nets have come fast on an obstruction. (Again, minor interest.)

ROMEO   Has no meaning in the Code.

SIERRA   I am operating astern propulsion. (As for Echo and India.)

X-RAY   Stop carrying out your intentions and watch for my signals. (As for Lima and Kilo.)

ZULU   I require a Tug. (Can you afford it?)

Here are a few suggestions for memorising the International Code.

ALFA   (Divers) Aqualung. Blue water, white sky.
BRAVO   (Danger) Bangs, Blood.
CHARLIE   (Yes) Si.
DELTA   (NUC) Difficulty.
ECHO   (Starboard) Bounces RIGHT back.
FOXTROT   (Disabled) Dancing.
GOLF   (Require Pilot) Guide me, Grid.
HOTEL   (Pilot) *Have* a Pilot.
INDIA   (Port) Black Eye.
JULIET   (Fire) Romeo found her hot stuff.
KILO   (Communicate) Kommunicate.
LIMA   (Stop) *Leave* off.
MIKE   (Vessel stopped) MIKING no way.
NOVEMBER (Negative.) Checkmate. End it. No.
OSCAR   (Overboard) Oblique.

PAPA	(Persons report) Blue *P*eter.
QUEBEC	(Quarantine) Yellow Jack.
SIERRA	(Astern) A*S*tern. The reverse of Papa. (PS)
TANGO	(Pair trawling) Tricolour.
UNIFORM	(Danger) *U* are standing into danger.
VICTOR	(Distress) 2 Vs, one inverted.
WHISKY	(Medical assistance) Red whisky in white glass with blue ice.
X-RAY	(Stop what you're doing) Cease. Cross.
YANKEE	(Dragging anchor) Yankee Ankor. A cable dragging at 45 degrees.
ZULU	(Tug) Z*U*L*U* T*UG*

Romeo is omitted because it has no meaning in the Code. It can be used as a sound signal in fog to warn of collision when stopped.

It is worth noting that red appears in Bravo, Foxtrot, Oscar, Tango, Uniform, Whisky and Yankee, all signals which warn of danger or urgency. The best way to learn the signals is by making your own flashcards (although they can be bought) using felt-tips and adding the meanings and morse equivalent on the back. Get someone to flash them at you and correct your mistakes.

The Morse Code

There are several ways of mastering the Morse Code, and the golden rule is practice. However, it helps to get off to a rapid start and the following method allows you to learn half the alphabet with speed. First, learn the letters which are all dots.

. ECHO .. INDIA ... SIERRA .... HOTEL

There are few words you can make with those four. HE, SHE, HIS and SHIE are about the sum total, so continue with the letters made up of all dashes.

— TANGO — — MIKE — — — OSCAR

Now we can make several sentences

THIS IS THE TIME.
SEE HIM SIT HIS TEST.
HE HAS A TEAM MATE.
SHE SMOTE HIM.

Next, learn two more, . — ALFA, and its mirror image — .
NOVEMBER.
This allows us to construct the following

SHE MOANS TO THE MATE. SHE HAS THE SAME NAME AS HIM.
SHE HAS MEAT AT HOME. HE HAS TO SIT HIS TEST.
SHE HAS SENT HIM INTO THE MIST. HE HAS TO MAINTAIN HIS HOME.

We now have nine letters of the alphabet. Another group can
be learned by looking for the three-cypher combinations and
spotting mirror images like

— . — KILO and . — . ROMEO (King's Regulations).

Now you could send

SEE THE HARE NEAR THE RAKE. THERE ARE THREE SHARKS THERE.
HE STARTS TO TEAR HIS HAIR. TAKE A ROSE TO HER MOTHER.
IS THAT A MARK ON THE SHORE? THE MARE HAS A RASH ON HER
EAR.
THEIR TEARS ARE TO SHAME THE MATE. IS THIS THE KEEN SEAMAN?
HE HAS TAR IN HIS HAIR. HE MEANS TO SEE THE MATE.
HE IS TO MAKE THEIR SHOES SHINE.

Four more are

. . — UNIFORM and — . . DELTA (United Dairies)
— — . GOLF and . — — WHISKY (Great Western)

IS HE TO GET OUT? I WISH HIM A GOOD RIDE. HAS HE A NEW GUN?
WHAT HAS HE DONE? GO TO THE WEST SHORE. GET ME SOME GOOD
STRING.
DID SHE GO WITH THE MATE? UNDO THE KNOT AND STEAM ON. DO
THE MEN GO TO SEA?
WHEN DOES SHE GO HOME? WHERE IS HIS NAME WRITTEN?

Out of the twenty-six letters you now have fifteen mastered.
Now start on the four-cypher letters.

—...BRAVO and ...—VICTOR (Beethoven's Vth).

Practise them with:

HE IS A BRAVE MAN. TAKE A BASKET HOME. SEE THE RAVEN IN THE TREE.

THIS IS A BARE ARM. SHE MEANS TO SAVE THE MAN. HE IS RATHER VAIN.

IS HE AT THE BANK? THAT BOAT IS THE BEAVER. HE WILL RAVE OVER THE BOAT.

The next four are:

.—..LIMA and ..—.FOXTROT (Low Frequency)

—.——YANKEE and ——.—QUEBEC (Why Queue?)

Practice phrases are:

WILL YOU FIRE THE GUN? IT WOULD BE QUITE AWFUL. DO YOU THINK THE QUEEN WILL COME? IS THIS THE WAY THE QUAIL FLY? WHO WILL MAKE THE BOYS QUIET? DID SHE FIND THE YOUNG GIRL? I LIKE YOUR QUICK FIRING GUN.

FILL YOUR BAG WITH THE BEANS. DOES HE ALWAYS DO IT LIKE THIS?

Next, .——.PAPA and —..—X-RAY (PX is the American Naafi.)

PLEASE FIX MY RADIO. WILL YOU PASS ME A SKEIN OF ROPE? GO TO THE QUAY AND MIX SOME PAINT. YOU MUST PAY THE TAX MAN. GO TO THE BAR AND GET A PINT. YOU MUST RELAX AND BE HAPPY. FIND ME A NEW BOX FOR MY PIPE. I THINK HE HAS SMALL POX.

That only leaves three odd ones .———JULIET, —.—.CHARLIE and ZULU, ——..

CAN YOU JOIN ME THIS AFTERNOON? HE IS IN A DAZE OVER THE JAM TARTS. MY COUSIN'S NAME IS JIM. WHERE ARE THE DOZEN CAGE BIRDS? PACK MY BOX WITH FIVE DOZEN LIQUOR JUGS.

Numbers are simple. All numbers have five characters and up to five the number of dots equals the number itself, hence:

1   .————

2   ..———

3 ...——
4 ....—
5 .....

At six the dashes reappear and to arrive at the number just subtract the dots from ten.

6 —....
7 ——...
8 ———..
9 ————.
0 —————

If you intend to send and receive more than a very simple message in morse you must learn some procedures which boil down to four essentials.

CALL UP SIGN AA AA (. — . —)
END OF MESSAGE AR (. — . — .)
FULL STOP AAA (. — . — . —)
RECEIVED T (—)

Examples

AA AA Please contact customs officer on arrival in port. AAA he has an important message for you. AR

AA AA Fishing vessel BM 629 reported sinking in position 50° 12′ N, 05° 47′ W. AAA Please render assistance. AAA Two Helicopters on their way. AR

AA AA I have picked up 2 crew members from French yacht Aventir. AAA They are suffering from exposure AAA Please have doctor waiting when I arrive Custom House Quay AAA ETA 1730 AR

EXERCISE THIRTY

(a) What is the meaning of a blue and white flag with a swallowtail?

(b) What is the meaning of a flag with three horizontal bars, yellow, blue and yellow?

(c) What morse letters would you use to:
(i) Call up a station?   (ii) Indicate end of message?
(iii) Indicate message received?

13 Radar

The average yachtsman tends to view radar in two conflicting ways. The majority would love to have it, especially in fog or at night, but recognise that it consumes an inordinate amount of electrical power, (50 to 60 watts are not uncommon) and to be of any use, must be mounted high up. With the average scanner weighing around 10 to 20 kilogrammes, stability is reduced. If you mount the scanner on top of your coach roof then its effective range is only 2.5 miles or the same as the human eye. Mounted 6 metres up on your mast the scanner can 'see' 5.1 miles, but its weight is now a problem. However, if the target is also 6 metres, then its effective range is doubled to ten miles which is very useful.

A radar is only as good as its operator. Liken it to the white stick carried by a blind man. It's unable to distinguish red from green, can from cone or rocks from buoys. Of course, there are highly sophisticated radars which can give you a tremendous amount of information. Go to the control tower at Heathrow and there you will see displays that, against the aircraft's echo, print its flight number, speed, course, altitude and practically what the captain had for breakfast. Some marine radars are becoming nearly as smart, but not, I think, in the average week-end sailor's price bracket.

The operator, to begin with, must know to what extent he can rely on his set and this depends on:

The weather and sea state.
The height of the scanner.
The reflective properties of the target. **Fig 13.1**

A lumpy sea will cause sea echoes, or clutter, a rain squall may obscure a target altogether, and a small yacht without a radar reflector will return an intermittent echo because fabric has poor reflective properties and the angle of the sail to the radar beam is constantly changing.

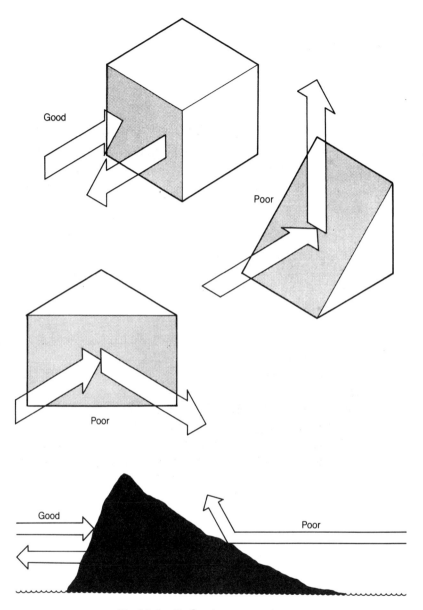

Fig 13.1 Reflective properties

Interpreting the Display

The beginner, when he first peers at the display, expects to see only the echoes of other craft, buoys, and the coastline. In fact, he usually sees a great deal more and not unnaturally, wonders how on earth he is expected to decypher all this extraneous information. **Fig 13.2** gives one some idea of what a typical display might show, and here is a brief explanation of each phenomenon.

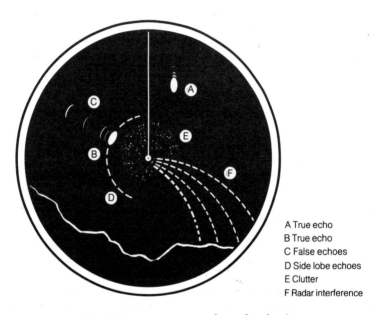

A True echo
B True echo
C False echoes
D Side lobe echoes
E Clutter
F Radar interference

Fig 13.2 Interpreting the radar display

1 **Shadow Sector** This is when the scanner's view is interrupted by a mast or some other form of superstructure. Its extent can be determined by turning your craft through 360 degrees when in the vicinity of a buoy and noting the bearings when it vanishes and re-appears.

2 **Multiple echoes** A succession of false echoes of a target caused by radar energy bouncing back and forth between

your scanner and the target. The nearest echo is the true range of the target.

3 **Side-lobe echoes** Although a radar beam is transmitted over a very narrow arc (around 1 degree), some energy is spread out on either side. Large vessels close by may return echoes from these side-lobes which appear as a series of dashes in an arc up to ninety degrees either side of the true echo.

4 **Clutter** (Sea/Rain) Waves close to the vessel, or heavy rain will return strong echoes usually in the form of a circle around the centre point of the display. Normally there is an adjustment to cut down sea clutter, but this should be used with care, as by so doing, one may reduce the echo of, say, a small yacht which is dangerously close to your vessel.

5 **Radar interference** Other vessels using the same frequency as your set may cause a pattern of dotted lines to appear on your display.

As a general rule of thumb, if an echo changes shape or position during each sweep of the scanner it is some form of interference. An echo which remains constant and which leaves a ghost image as the sweep continues is almost certainly a true target.

The Movement of Echoes

The most important thing for a radar user to remember is that the track of a target's echo on the display does not indicate the course of the target. The only time when it does is when your vessel is stopped, and at all other times the movement of echoes only shows a RELATIVE course. **Fig 13.3**

I am assuming here that you are not the proud possessor of a True Motion Display set, into which you feed your course and speed, in which case the echoes of stationary targets and land remain stationary on the screen while your position is made to move across the screen and the echoes of targets making way will indicate their actual course and speed. These are normally only available to be used in conjunction with gyro compasses.

On an ordinary yacht radar, when your vessel is moving but

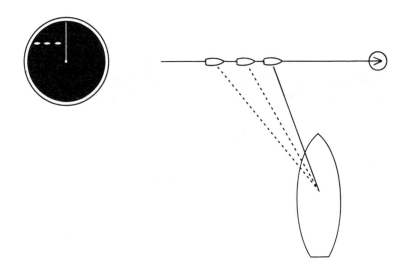

Fig 13.3   Your vessel stopped

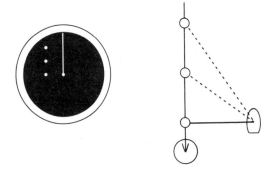

Fig 13.4   A stationary target

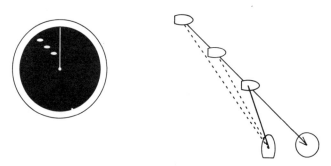

Fig 13.5   Both vessels moving

the target is stationary (eg a buoy) its echo will appear to move in a parallel but opposite direction to your course. **Fig 13.4**

When both you and the target are making way the track of the echo is a combination of both vessels' movements. **Fig 13.5**

Closest Point of Approach (CPA)

It is vital to be able to forecast how close a target is likely to approach one's own position and the time at which it will do so. To do this take a chinagraph pencil and plot the successive positions of the target's echo. The line thus produced is called the Apparent Motion Line and it will determine her closest point of approach and whether avoiding action on your part is necessary.

In **Fig 13.6** we plot a target bearing 310° range 2.9 miles, Time 1200. At 1205 the echo bears 320° 1.8 miles. At 1210 it bears 345° 1.2 miles. What is the target's CPA, time of CPA and at what time will she pass ahead?

Using a chinagraph we can plot ahead and find she will cross

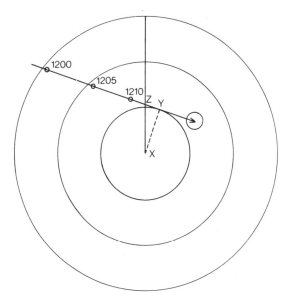

Fig 13.6 Closest point of approach

our path at Z and be closest to us at Y. By measuring the Apparent Motion Line from the 1200 to 1210 positions we can deduce that her RELATIVE speed is 12 knots (2 miles in 10 minutes), and that she will pass ahead of you by 1.1 mile(Z) at 1211.5 and her CPA will be 1 mile at 1213 (Y). **Fig 13.6**

Collision Courses

If the apparent motion line passes through your own position at the centre of the display then the risk of collision is very high. If your set possesses a cursor this can be rotated to pass through the target's echo when first sighted. If, after a few minutes, the echo remains on the cursor line then theoretically the echo will pass through your position and so will the target. If, on the other hand the echo moves right or left of the cursor, she is drawing ahead or astern, and the risk is minimised. A close eye should be kept on the target however, until it starts to distance itself from you. In the case of an overtaking vessel remember to maintain your course and speed until she is well past.

Example
1145 target bears 130° 8.4 miles
1150 target bears 130° 6.4 miles
1155 target bears 130° 4.4 miles

Use the upper compass rose on Chart 5067. Since these bearings are all relative to your ship's head it matters not whether you use the inner magnetic rose or the outer. I recommend the latter, and for our purpose we assume we are in an unstabilised or ship's head up mode.

From the rose's centre, which is our position, draw a line through 130° and mark off 8.4, 6.4 and 4.4 miles, timing them 45, 50 and 55 respectively. Projecting the time forward you will discover that a collision will occur at 1207 unless avoiding action is taken. Who should take it? Answer, the other vessel as she is in your overtaking arc. Had the bearing been 112° or less then she would have the right of way and your duty would be to give way. A classic case where slowing down and allowing her to pass ahead would probably be the best course of action.

EXERCISE THIRTY-ONE
(a) 1030 target bears 240° 10.0 miles
     1040 target bears 245° 8.0 miles
     1050 target bears 253° 6.1 miles
     What is the CPA and its time?
(b) 1300 target bears 066° 8.0 miles
     1310 target bears 049° 6.0 miles
     1320 target bears 021° 5.0 miles
     What is the CPA and its time?
(c) 0330 target bears 180° 9.5 miles
     0335 target bears 183° 8.0 miles
     0340 target bears 187° 6.7 miles
     What is the CPA and its time?

Radar Assisted Collisions

Strange as it seems they happen and despite frequent articles on the subject in the yachting press they continue to happen far too frequently. The most celebrated example was probably that between the *Stockholm* and the *Andrea Doria* off New York. They were both modern, well-found ships (the *Andrea Doria* was the top Italian passenger liner) run by competent professionals and although in fog they were both aware of each other. There was some confusion to start with as the *Andrea Doria* assumed the *Stockholm* was the Nantucket Light Vessel. Had the Officer of the Watch had more time to check his radar plot he would have noticed that the 'light-vessel' was steaming along at 17 knots! In the final minutes they both realised the enormity of the situation, but one had now turned to port and was struck by the other altering to starboard. **Fig 13.7** shows how it could have happened.

Interpretation by A

A first sights B at a range of 9 miles right ahead. Correctly assuming an end-on situation, she starts to make a leisurely

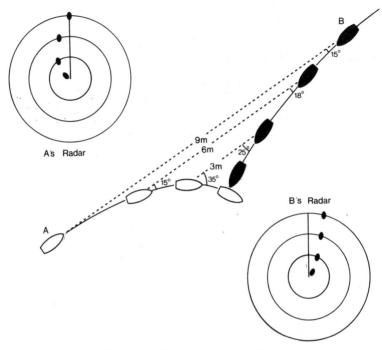

Fig 13.7 Radar assisted collision

alteration to starboard. At 6 miles B is 15 degrees on the port bow and as the starboard turn increases 35 degrees at 3 miles. At one mile it was obvious that the two vessels would not pass safely and an even greater alteration was finally made, but to no avail.

Interpretation by B

B sights A at 15 degrees on the starboard bow and assuming rightly that if both held their course and speed they could pass safely starboard to starboard, continued on course. At 6 miles the angle on the bow is still 18 degrees, but B decides to put a little more room between them and makes a small alteration to port. At 3 miles the angle has opened to 25 degrees but again at a mile, the range had closed rapidly and disaster ensued.

Remember if both were steaming at 15 knots their approach speed would be 30 knots and they would cover that last fateful mile in two minutes.

The situation might have been avoided if the following precautions had been taken.

Neither vessel had turned to port.
Both had reduced speed or stopped at 1 mile.
Either had taken substantial action in the early stages. 'A' should have made a 60 degree alteration at 3 miles, instead of small changes of course.

Plotting

The easiest way out of the problem is by plotting the other vessel's TRUE course (as opposed to relative motion) and speed. Once this is known there should be no cause for the situation described above.

Imagine that when the target is first sighted it is stationary, or is a buoy. It is simple to estimate the relative position of such an object in the future because the apparent track of a buoy will be parallel and opposite to one's own course and its apparent speed will be your speed.

Choose a suitable time interval. Six minutes is useful as it is one-tenth of an hour and draw a line with a chinagraph pencil from the first echo parallel with and reciprocal to your course, its length being your speed over that period ($\frac{1}{10}$ of your speed). In other words plot the position of the imaginary buoy six minutes later. Six minutes after the first sighting plot the position of the target. The difference between the two must be due to the vessel's own movement. By joining the two you will immediately perceive the target's course, and if you measure the length of the line and multiply it by ten you will calculate its speed. By this method, one can avoid the situation described above.

For example in **Fig 13.8** a target bears 330° range 5 miles at 1000. You are steaming 000° at 10 knots.

Plot target echo and draw a line 1 mile long parallel to your course.
Name it point A.
1006 Plot position of target echo (B) 350° range 3 miles.

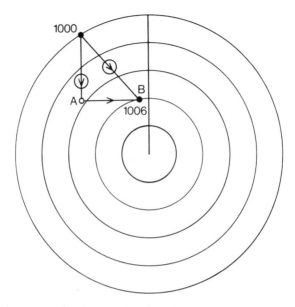

Fig 13.8 Plotting another vessel's true course and speed

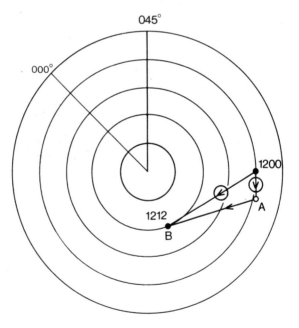

Fig 13.9 Plotting another vessel's true course and speed

Join AB. AB is vessel's course and length × 10 is its speed. Target's course 095° speed 20 knots.

In **Fig 13.9** the target bears 090° relative, distant 4 miles at 1200. Your course is 045° at 5 knots. At 1212 target bears 160° relative at 2 miles. What is target's true course and speed?

Plot target's 1200 echo and draw a line 1 mile long (⅕ of 5 knots) and reciprocal to your course, ie 225°.
1212 Plot position of target echo (B).
Join AB. AB is vessel's course and its length multiplied by 5 is target's speed. Target's course 300°. Speed 16.5 knots (3.5 × 5).

Fixing by Radar

A long range fix cannot be regarded as reliable as radar bearings are subject to error and it is always possible to mistake high ground inshore for the actual coast as **Fig 13.10** illustrates.

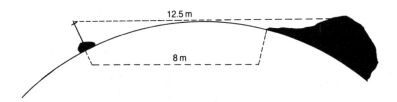

Fig 13.10   A false range

Wherever possible supplement the range with a VISUAL bearing which is likely to be more accurate, but if you are using two points of land a double check with both ranges and bearings will at least show whether you have identified the correct targets.

In **Fig 13.11** if high point B had been mistaken for low-lying A the fix by ranges would be inside the vessel's position and the fix by bearings outside.

X is a false fix by ranges and Y a false fix by bearings. Z is a

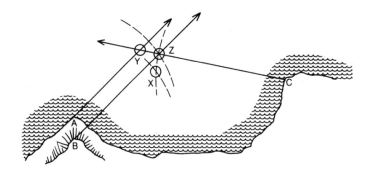

Fig 13.11 False fixes

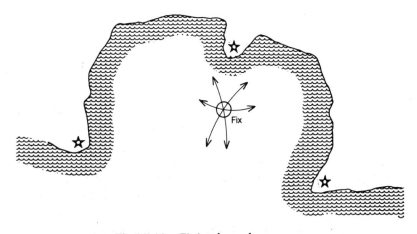

Fig 13.12 Fixing by radar ranges

correct fix. C is correct in all cases, but B has been mistaken for
A.

Radar ranges can reliably be used where three identified
targets are available. As with a visual three point fix they
should be fairly widely separated as in **Fig 13.12**.

Cross Index Ranges

Suppose you wish to keep two miles clear of the coastline or a
series of headlands all you need do is set up the 2 mile range
ring or strobe, and draw a line on the display two miles from its
centre. As long as the echo of the land remains outside the line

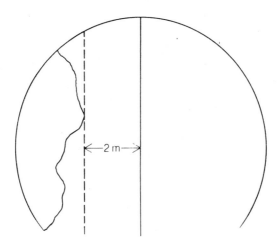

Fig 13.13   Cross index range

you will clear it by at least two miles. **Fig 13.13**

But beware of using this method on a single target, for example, if you wish to pass 2 miles clear of a light vessel marking a reef and there is a current on the beam of two knots. Using the above method it would appear that in **Fig 13.14** you will leave the light-vessel two miles on the starboard beam and well clear of the reef. But in **Fig 13.15** we see what is actually happening.

The correct way to tackle this problem is as follows:

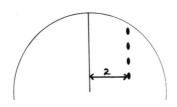

Fig 13.14   Clearing a single object

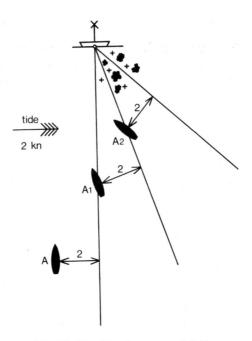

Fig 13.15 Plotting set and drift

Fig 13.16 shows a light-vessel on the starboard bow, first detected at A. If there were no tidal stream the light-vessel would later have appeared at B instead of C. As the light-vessel is stationary BC must represent the set and drift of the tide in that interval.

In order to pass clear of the light-vessel at 2 miles you must do a little plotting.

> With centre your vessel (at centre of compass rose) draw a circle radius two miles.
> Decide on a time to alter course to port and estimate the light-vessel's position at that time (D).
> From D drop a tangent down to the circle of closest approach.
> Draw CE parallel to your tangent.
> Plot B (the position the light-vessel would have reached had there been no tidal set) and join CB.
> With centre B and radius AB draw an arc to cut CE at X.
> The angle ABX is the amount the course must be altered.

Now I appreciate that this is big-ship stuff and there are many difficulties for the small boatman, not least of which is the liveliness of the boat, its tendency to yaw, poor radar range and lack of facilities generally, but it is vital that the mariner should be aware of problems such as the above and know how to rectify them.

EXERCISE THIRTY-TWO
(a) 1100 Cape Cornwall 2.95 miles. Longships Lt 2.35 miles. Land's End 2.95 miles. What is marked on your chart beneath your 1100 position?
(b) 1530 St Anthony Hd Lt 3.5 miles. Manacle Pt 2.9 miles. Rosemullion Hd 3.5 miles.
    What is lat and long of your 1530 Fix?
(c) 1630 Dodman Pt 5 miles. Nare Hd 4 miles. St Anthony Hd Lt 6.4 miles.
    What is lat and long of your 1630 Fix?

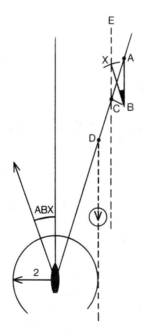

Fig 13.16 Plotting to allow for set and drift

Anchoring in Fog

Let us assume that you wish to anchor due south of a buoy. In this case there is a prominent rock due north of the buoy which provides a useful transit.

On the display and using a chinagraph pencil draw a line commencing at the centre spot and superimposed over your heading marker. Mark off 1 to 5 cables on this line, plus a 'Let Go' mark 1 cable from the centre spot. As you approach the buoy (keeping it in transit with the rock) it will appear to move down your heading marker toward the Let Go mark. At two cables you will go dead slow and perhaps stop at one cable. When the buoy reaches the Let Go mark you anchor. **Fig 13.17**

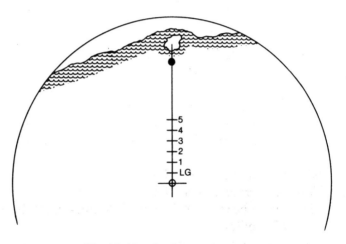

Fig 13.17   Anchoring by radar

Radar Responders (Racon)

Certain lighthouses and light-vessels, particularly those near congested shipping lanes, identify themselves by responding to the electrical energy transmitted by your scanner. This appears on your display as a thick bar extending away from the target. It does not appear with every sweep of the scanner but usually every dozen or so in order not to obscure targets which lie in its path. These stations are marked on the chart (eg Bishop Rock, Seven Stones, Channel Lightvessel).

Another type of radar responder is a small hand-held receiver such as the Lo-kata 'Watchman'. This will provide you with an audible and visual warning and a bearing of any vessel, large or small, operating a radar within five miles of you. Clearly designed with the single-handed sailor in mind, it is an obvious boon when crossing busy shipping lanes. Its current consumption is minimal and they cost about the same as a small colour television.

EXERCISE THIRTY-THREE
For the following use the True rose on Chart 5051 bearing in mind that all bearings are relative to your ship's head. In (a) you are steaming 190° and a relative bearing 300° would be a true bearing of 130°. (190° + 300° − 360° = 130°)

(a) You are steering 190° at 12 knots.
     1145 target bears 300° R range 8.4 miles. 1155 bearing
     300° R range 4.4 miles.
     What is its true course and speed?
(b) You are steering 100° at 6 knots.
     1300 target bears 326° R range 5 miles. 1320 bearing 280°
     R range 8 miles.
     What is its true course and speed?
(c) You are steering 210° at 18 knots.
     0330 target bears 330° R range 9.5 miles. 0340 bearing
     334° R 7.5 miles.
     What is its course and speed?

14 Radio Aids

During World War II both sides developed directional radio beams to assist their bombers to locate their targets. The war over, the same principle was applied to civil uses and very soon small, efficient radio direction-finders were on the market at a reasonable cost. In the 1960s the USA began putting satellites into space, again for military use. In recent years, however, the advance in microchip technology has put receivers for this type of information on to the small boat market though they are still fairly expensive.

Most of us are familiar with the fact that a radio with a ferrite rod aerial mounted longitudinally achieves its best reception when placed at right angles to the transmitted beam. Conversely, reception is weakest when the aerial points to the beacon. Radio DF uses this principle, and it is possible to construct a cheap DF set by mounting a portable radio on a pelorus or sectored board, which is free to rotate. Most cheap radios can receive LW transmissions between 285 and 312 kHz, and all one does is line the pelorus plus radio up with the fore and aft line of the boat and you can perceive relative bearings of the transmitting beacon.

However, as in all else you get what you pay for and DF sets like Seafix start at the price of a couple of good dinners. A mid range instrument like the Lo-kata costs about the same as a colour TV and this includes such refinements as a light which fades as the null is found and can be adjusted to give a very precise bearing. It also incorporates a receiver tuned to 200 kHz. For the price of a videocamera you can buy a semi-automatic set. Headphones are better than loudspeakers and for the beginner, digital tuning is better than tuning by hand.

Method

First plot your DR position. Since no matter how accurate you are you're going to get two nulls, reciprocals of each other, you need to know where you are relative to a station. For instance, if you are east or west of the Lizard or north or south of the Eddystone. Incidentally the latter only transmits in fog and then at a range of only twenty miles, but that is when you need it most.

Next, tune to the chosen frequency which you will find in the Almanac under Radio Beacons. Turn the set with its compass until the signal becomes null. By rotating the set either side of the null, eventually a reliable bearing, to within plus or minus 3 degrees will be obtained.

Note the bearing and carry out a similar procedure for another beacon. Finally, plot them on the chart.

Example In mid-Channel, South East of Start Point, you tune to the Channel West group of beacons which transmit on 298.8 kHz. Your first beacon is Start Point (call sign SP . . . . — — .) transmitting at 1, 7 and 13, etc minutes past the hour.

Identify the call sign which is repeated three to six times for twenty-two seconds. This is followed by a long dash which lasts for twenty-five seconds, during which time you find your null. There is a second identification period lasting eight seconds followed by five seconds silence.

For your second beacon you choose the Casquets (QS — — . — . . .) which transmits at 2, 8 and 14 etc minutes past the hour and repeat the process.

This is the working. We will assume that you have corrected your course to one of 073° True.

START POINT	CASQUETS
D/F bearing 259° rel	022° rel
Quad error −1°	−1°
Relative bearing 258°	021°
+ 073°	+ 073°
True bearing 331°	094°

Of course, there is nothing to prevent you from changing frequency if you think you could obtain a better cut. For instance off the Wolf Rock, having got a reasonable bearing of the Lizard at 298.8 kHz, you could switch to 308.0 kHz and receive Round Island beacon.

In the example worked above you will have noticed a line labelled 'Quad Error'. This is explained below.

Errors

1 Quadrantal errors. Radio waves are distorted by standing rigging, guard rails and metal masts so when in sight of a radio beacon a card should be made up comparing the radio bearing with a visual sighting, thus:

A	B	C	D	E
Mag hdg	Mag bg of Bn	Rel bg (B – A)	D/F bg (rel)	Error (C – D)
085°M	128°M	043°	039°	+4°
070°M	130°M	060°	057°	+3°
057°M	138°M	081°	079°	+2°

and so on.

The result will be a DF 'Deviation card' for all ship's headings.

2 A vessel yawing or rolling will cause relatively inaccurate bearings to be taken.

3 The accuracy of a bearing decreases with the distance from the beacon. Any bearings received when the vessel is at the limits of the beacon's range should be treated with caution.

4 Night effect. From one hour before sunset to one hour after sunrise radio signals tend to be reflected more strongly from the ionosphere than the more usual ground waves and an unsteady null is obtained. During the above period DF bearings should only be taken of beacons which are within twenty-five miles of the vessel.

5 Coastal refraction. Waves travelling along a coastline or parallel to it will be deflected either towards or away from it, and will not indicate the direction of the beacon. To

avoid this, either use marine beacons or those whose beam has crossed the coast at right angles or nearly so.

There is not a lot you can do about 2, 3 and 4 except to try and avoid taking bearings when these conditions prevail. If it is necessary to do so, they must be treated with caution. Remember, if using three bearings you are almost certainly going to get a cocked hat. Out in mid-Channel a triangle large enough to cause concern inshore is not dangerous, but checks should be made to reduce its size. It is the custom to draw a circle around a cocked hat indicating that your position is somewhere within that circle. As soon as possible verify your position, either by radar or visually.

Loran, Decca, Satnav

The serious mariner with a deep pocket and a beefy generator may like to consider one or more of the above sophisticated devices. They have two things in common. They are expensive and they require comparatively massive power supplies, (8–25 Watts). The cheaper versions also require charts overprinted with their appropriate lattices.

Loran, which means long range, operates in the North Atlantic, North America and the Mediterranean sea. It has a range of 1,500 miles and a claimed accuracy of ± 1 cable. You need lattice charts for the cheaper versions which cost around the same as a colour TV while three times that buys you a set with a digital readout of your lat and long. They consume about 15 watts.

The Decca chain of transmitting stations has been established round our coasts since World War II. It now extends from North Cape to Gibraltar and has a range of from 175 to over 300 miles. It can be accurate to within twenty-five metres and the cheaper versions need latticed charts for their interpretation. The latest models provide a digital readout of your lat and long, every twenty seconds, consume about 20 watts and cost about the same as a video-camera.

Satnavs by Walker or Decca work on an entirely different

principle from the others. They utilise the United States Navigation Satellite system, the installation consisting of a small dish aerial and a computer. Satellites orbiting eighty miles above the earth transmit their orbital positions, and as they pass close to your receiver the computer calculates your lat and long which is displayed as a digital readout. They require no special charts and as they are computers they will accept inputs such as waypoints (or intended DR positions), tidal stream, speed and so on.

By comparing your position with your waypoint they will predict your new course to steer and update your ETA. Apart from cost which is similar to Decca, their main disadvantage appears to be that quite large periods of time must elapse between satellite passes, but as more satellites become available this interval must surely decrease. The Walker only consumes 8 watts.

In addition to these hefty devices, the navigator should not ignore the pocket calculator. For astro-navigation there is on the market for well under £100 a calculator which, having received an input of your DR Lat and Long, time and altitude, will compute azimuth and intercepts for sun, moon and star sights, thus taking all the drudgery out of it. In fact microcomputer technology is developing so fast that books on the subject can become obsolete before they are off the presses.

EXERCISE THIRTY-FOUR

1 Using an Almanac find the frequencies and call signs of

    (a) North Foreland   (b) Dungeness   (c) Nab Tower
    (d) St Malo.

2 Name four errors to which DF bearings are subject.
3 What is the main difference in principle between a Decca and a Satellite navigation instrument?
4 Apart from initial cost, what two disadvantages do these sophisticated radio aids possess?
5 Why would Oessant SW, St Helier Harbour and Poole Harbour beacons be of little use to you in mid-Channel? (Check in Almanac.)

15 Passage Planning

It would be foolhardy for the week-end sailor whose experience has been limited to pottering around in estuaries to suddenly attempt an offshore passage. Much better that he makes his first voyage within sight of land and to this end the following conditions should apply.

1 The passage can be completed in daylight.
2 Departure should be timed to give the longest period with a fair tidal stream.
3 The track should be about 1 to 2 miles offshore wherever the coast is steep-to, so that shore objects can be easily identified. Where the coast is shelving the average yacht drawing perhaps three metres should aim to pass outside the ten metre line.
4 The latest weather forecast has been obtained, either from the BBC's broadcasts or by telephoning one of the Met offices or the Coastguard. Reports of an approaching depression or of fog should never be ignored.
5 Finally, contact your local Coastguard, giving him all the particulars he asks for. Telephone him again on arrival, or if there has been any change of plan and you have returned to port, or found another destination. Search and Rescues are expensive.

Take out 5051 and plan a passage from Black Rock Bn (Falmouth) around the Lizard to Penzance. Assuming we have a motor-sailer, capable of five knots through the water, then passage time will be between six and seven hours, an ideal cruise for a summer's day. Wind is SW Force 4, so we should manage the trip on a close or broad reach, although in the lee of the Lizard it may be rather fluky.

Next, if we look at the Tidal stream table (bottom right) under columns J, E, F and D we see that the time to leave is around HW Devonport and the tide tables show that on Saturday May 3rd 1986, HW Devonport is 0733 GMT or 0833 BST. This is the time to be at Black Rock Bn giving us a civilised departure and a half-hour's motor down to the mouth of the Carrick Roads.

At the beacon we must set a course to make good the yellow buoy 2.4 miles off Nare Hd and proceed to leave the Manacles buoy ½ mile to starboard. The tidal stream at this time is 040°, rate 0.4 knot (neaps) so it is negligible. Off Black Hd the tide is slack and as we approach the Lizard it is setting 203°, 0.2 knot. We must alter course to round the Lizard, aiming to stay out of the race as it is wind against tide.

Making sure we stay outside the Boa we come up towards Penzance with wind on the beam. We are now in Area C at six and seven hours after HW Devonport, which gives us a useful lift of 0.5 knots at 280°. We must now plot this track on the chart. For this exercise deviation has been omitted.

The Navigator's Notebook

Once the plan has been put on the chart the courses, distances, times, etc must be transferred to a notebook so they can be referred to easily. For this passage the page might well look like this.

Date 3/5/86   HW Devonport 0833 BST   Variation 8° W

0800   Slip Mylor Yacht basin. Co as required for Black Rk Bn.
0830   Black Rock Bn abeam. A/C 180° M.
0915   Yellow buoy abeam.
0942   Manacles buoy abeam 6 cables. A/C 216° M. Start engine.
1030   Black Hd abeam 1.8 miles.
1130   Lizard 316° M. A/C 278°. Stop engine. Tidal stream 203° 0.2 knots.
1230   Lizard 053° M. A/C to make good 331° M. Steer 340° M.

1440   Mountamopus abeam 1.4 miles.
1520   ETA Penzance.

As you progress, fixing as you go, you can discover whether you are inside or outside your track, ahead of or behind your ETA. For instance, crossing Mount's Bay you may find leeway is cancelling out tide effect and it might be better to steer 331° in order to prevent yourself being pushed up towards St Michael's Mt.

Incidentally you will have noticed that I wrote 'Manacles Buoy abeam 6 cables' in the plan when in an earlier chapter I stated clearly that you should never fix on a buoy. However this is not a fix but a point to aim at. When one arrives off the buoy one should fix on surer marks such as Nare Hd, Black Hd and St Keverne Church Spire.

Paradoxically planning a cross-Channel passage is simpler. Assuming that the crossing will take about twelve hours then the stream will flood and ebb for about five hours each way with two periods of slack water. Since it is largely on the beam there is no need to make allowance for speed made good, nor is it necessary to construct twelve vector triangles. A single course steered across both flood and ebb will achieve the same result, although it will divert you from your intended track. While this is safe enough in the western Channel it is inadvisable to use this method crossing from the east, such as from Ostend to Ramsgate, as you could easily be swept on to the Goodwin Sands.

Example A motor yacht sailing from Dartmouth to Guernsey sets a course from two miles off Start Pt. The distance to Les Hanois is 60 miles and at 5 knots the passage will take twelve hours and she will be in areas H and J throughout. Tidal streams for those areas are shown as follows:

Hours	Set	Rate	Hours	Set	Rate
1	037°	0.7 knots	7	252°	1.1 knots
2	041°	1.3 knots	8	239°	1.3 knots
3	048°	1.5 knots	9	230°	1.4 knots
4	051°	1.5 knots	10	221°	1.3 knots
5	046°	1.0 knots	11	206°	0.8 knots
6	057°	0.3 knots	12	115°	0.4 knots

The course to make good from Start Pt to Les Hanois is 150° M and if we steer that the ebb and flood will cancel each other out. The fact that there is seven hours of flood compared with five hours of ebb can be discounted as two of the hours (sixth and twelfth) only add up to 0.7 of a knot. The result will appear as in **Fig. 15.1**.

But what if the passage takes longer or less than twelve hours? Then simply calculate the balance or remainder after the streams have been equalised and find a set and rate which can be applied to produce a big vector triangle. For example, **Fig 15.2** shows a fifteen hour passage for the same voyage. For simplicity's sake, the sets have been averaged.

1	045°	0.7 knots	8	230°	1.3 knots
2	045°	1.3 knots	9	230°	1.4 knots
3	045°	1.5 knots	10	230°	1.3 knots
4	045°	1.5 knots	11	230°	0.8 knots
5	045°	1.0 knots	12	087°	0.4 knots
6	045°	0.3 knots	13	087°	1.0 knots
7	230°	1.1 knots	14	087°	1.2 knots

A   Total drift at 045° = 6.3 miles
B   Total drift at 230° = 5.9 miles
C   Total drift at 087° = 2.6 miles
Balance (A + C − B) = 3.0 miles in direction 066° (average)

Method
Plot the balance of the tidal stream at point of departure.
Join end of tidal stream vector to end of track.
Allow for Variation, Deviation and Leeway.
That is your course to steer.

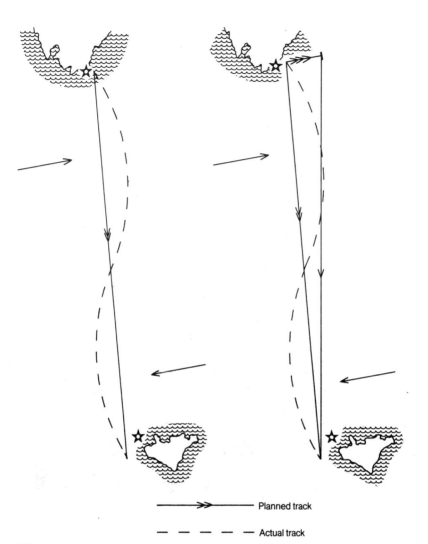

——————⟶⟶—————— Planned track

— — — — — — Actual track

Fig 15.1 Passage planning
with symmetrical tidal stream

Fig 15.2 Planning allowing
for an unequal set

Separation Zones

As soon as you examine the Passage Chart (2675) you will notice magenta rectilinear shapes in mid-Channel, off Land's End and the Isles of Scilly, and off Ile d'Oessant. These indicate the zones which separate the main channels used by ocean-going shipping entering or leaving the English Channel.

These lanes are between five and seven miles wide while the Separation Zone is between three and six miles wide. Additionally, there are Inshore Traffic lanes off Ushant and the Casquets for coastal traffic and you would be well advised to avoid them both when cruising.

It is incumbent on you when crossing the Channel to do so as swiftly as possible. You therefore plot a course to traverse the lanes at right angles or as nearly so as is practicable. To do otherwise is not only to break International Law, it is akin to jay-walking in a busy street. If you need to make an Easting or Westing either do so before joining the lanes or after leaving them. Do not run parallel with them in the Separation Zone which may well contain its quota of fishermen.

Both lanes and Separation Zones provide a useful navigational aid. Make a note of the times you expect to arrive and depart each lane and add them to your expected times for lights, etc. Crossing the lanes will confirm your DR or otherwise!

Crossing the lanes at right angles or nearly so is even more important in the Inshore Lane when you may be clearing a headland and stumble on a large coaster (or even an ocean-going monster in ballast) at very short notice. For example, one might be tempted to depart the Runnelstone for the Scillies on a direct course of 260°. But large vessels leave the down lane off the Lizard to round Land's End inside the Wolf Rock en route for South Wales on a course of 300°, a difference of only 40 degrees. Conversely a vessel destined for Rotterdam will come round the Longships on 120°, while you, returning from the Scillies might be on 080°. Far better to set a course for the Wolf Rock on, say, 235°, making an angle of 65 degrees, and then come round to 265° for St Mary's.

The Deck Log

Whilst all this is going on it is essential to keep a record of the passage, in other words to log it. Jotted notes on scraps of paper, or backs of envelopes soon become an illogical muddle and get lost or blown over the side. One can purchase expensive printed logs with sixteen column headings and at the other end of the scale you can buy an exercise book and rule it off into six columns. This is about the minimum necessary.

Time	Log	Course	Wind	Barometer	Remarks
0830	3.6	180° C	SW 3	1019	Black Rk
0912	7.4	180° C	SW 4	1019	Y. By
0939	9.5	180° C	SW 4	1020	Manacles 8 cab.

Making a Landfall

The end of a passage requires the greatest care of all. The ideal time to approach your haven, especially if unfamiliar, is at around first light. You can then positively identify your lights, buoys that would be obscure by day show clearly, and leading lights become obvious. Having made sure you know where and what everything is, fix your vessel and stand off, preparing to enter in daylight. If this is not possible or convenient make sure of your position before attempting to close the land.

Always approach a strange coast at right angles, with your echo sounder operating. From your largest scale chart work out your safe line of approach, identifying any transits. Check the stream which often runs counter to the main current when close inshore. The behaviour of buoys, pot markers and anchored vessels will provide clues as to its set and rate. Sketches and photographs in Admiralty Sailing Directions and the various Pilots, not to mention the Stanford Guides to the Channel Islands and Normandy and Britanny coasts, are invaluable and should be studied diligently while preparing the passage.

Nonetheless there is often anxiety about the identification of the actual entrance unless it has some very distinctive feature. There is no need to be anxious and many navigators worry about finding a river entrance long before there is any real

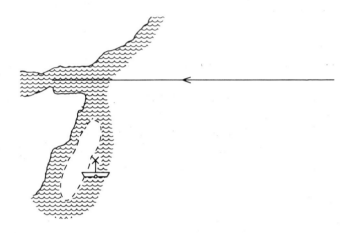

Fig 15.3   A risky landfall

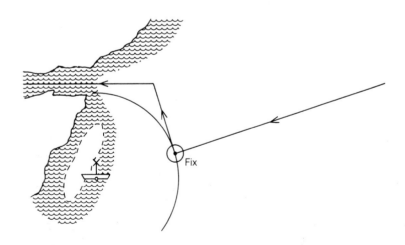

Fig 15.4   A safer method

opportunity of identifying it. Just so long as you can fix reliably and frequently, you can almost always approach the coast to within a mile when all should become clear. **Figs 15.3 and 15.4**

Finally, before entering make sure that there is sufficient water inside, that there is no bar or shoal to be negotiated and the port is showing entry signals, usually at night three vertical green lights.

EXERCISE THIRTY-FIVE
(a) List five precautions to be taken when planning a coastal passage.
(b) Name five pieces of information essential for the Deck Log.
(c) When is the ideal time to make a landfall and why?
(d) Why is it prudent to approach the land at right angles instead of running in along the coast?

16 Handling Under Power

Although this subject properly belongs in manuals of seamanship, it is of little avail knowing how to cross the Channel, avoid collisions and navigate at night without being able to make your way to a berth smoothly and efficiently once you have arrived. Therefore some words on handling under power at close quarters will not come amiss.

Paddle Effect

It is important to remember the three forces which act on a craft under power.

Thrust   The driving force of the propeller.
Rudder   The steering effect of the rudder when water is moving over its surface.
Paddle effect   The transverse force imparted by the propeller mainly through its lowest blade passing through denser water than the upper. Sometimes called the 'pull of the lower blade'.

This third one produces interesting effects so let us go into them in more detail. In a single-screw vessel where it is rotating clockwise (viewed from astern) this paddle effect is continually trying to force the stern to starboard and the bow to port, which results in most single-screw vessels carrying one or two spokes of starboard helm when going ahead. Conversely, when the engine is put astern the stern will be pulled to port. And another interesting fact comes to light. Although we appear to lose power when going astern (due to the inefficiency of a propeller designed to thrust ahead) the sideways swing to port is even more noticeable than its opposite when going ahead and this we can use to our advantage. **Fig 16.1**

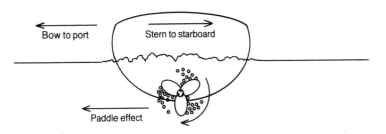

Fig 16.1 Paddle effect, single screw

Securing to a Buoy

Approach the buoy against the stream, but if tide is slack, against the wind. If wind is against stream still use the stream as your brake as it takes a very strong wind to overcome the effect of tidal stream or current. In a small yacht where the buoy can be reached by a crew member lying on the fo'c'sle, have your cable passed through a fair-lead and bring the buoy up to the bow. Pass the cable through the buoy shackle with plenty of slack and go slow astern. Bring the cable through the other fairlead, stop engine and secure. **Fig 16.2**

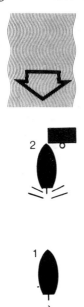

Fig 16.2 Approaching a buoy

With a large yacht or motor cruiser it may be easier to bring the buoy amidships, where the freeboard is lowest. Steer for the buoy, keeping it very fine on the starboard bow. Have the cable rove through the starboard fairlead, or better still, bullring if you have one, and brought back amidships on the starboard side. Stop engine as the buoy comes alongside the bow and it should drift down to where your crew is waiting. Go astern and the bow will swing towards the buoy bringing it within reach. Drop the cable through the shackle and then back to the bullring as the vessel makes sternway. Secure to bitts on the port side and stop.

Twin-screwed vessels with pronounced flare and rake to their bows may care to try what is known as the customs launch approach. For this, rig your most massive fenders across the transom at waterline level. Come to the buoy stern first, until it is hard against your fenders and secure it by two warps led out from both port and starboard quarters and passed back to their respective bitts.

Slipping a Buoy

Come slow ahead, let go the cable's end and hauling in through the bullring, clear the cable from the buoy. Go astern

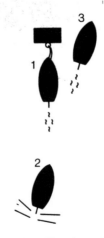

Fig 16.3   Clearing a buoy

and when clear of the buoy come ahead, leaving it to port because your bow will have swung to starboard by your kick astern. **Fig 16.3**

Turning Short Round

At this point you are clear of the buoy but perhaps facing upstream and you want to get out to sea. There is little room to manoeuvre so turning short round is the answer. Turning short round is something like a three-point turn executed on the road, and it should be done by turning to starboard. This starboard turn is made despite what we have learned about paddle effect trying to rotate the vessel to port and making the turning circle tighter to port. We turn to starboard because part of the turn involves going astern and this is where we want paddle effect to be greatest. It is best shown in **Fig 16.4**.

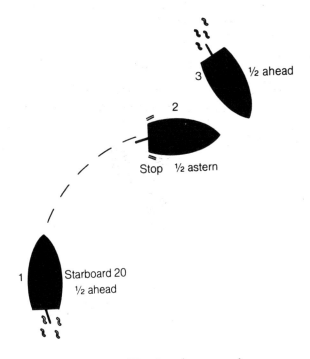

Fig 16.4 Turning short round

1 Starboard 20. Half ahead.
2 Stop engine. Half astern.
3 Stop engine. Half ahead. Midships.

If the wind is on the port beam it is worth giving a kick astern to start with, because the stern of a vessel going astern will always seek the wind. Once she is swinging, put the helm over, go ahead and give another kick astern before she gets too much way on. **Fig 16.5**

With the wind ahead it is very much a three-point turn, again using this wind-seeking capability of the stern. **Fig 16.6**

1 Starboard 20. Half ahead.
2 Midships. Stop engine.
3 Port 20. Half astern.
4 Midships. Stop engine. Half ahead.

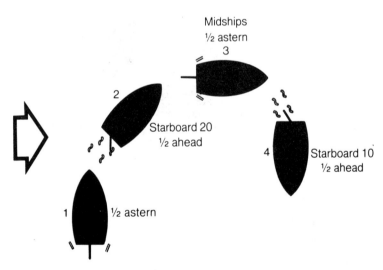

Fig 16.5   Turning short round (wind abeam)

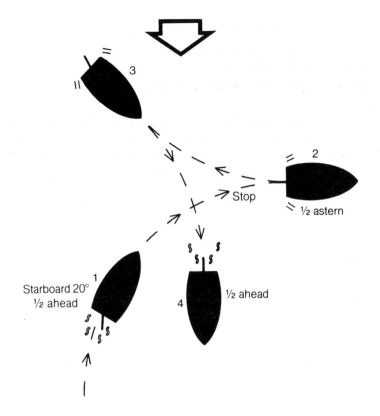

Fig 16.6   Turning short round (wind ahead)

Berthing

First attempts at coming alongside should ideally be made on a calm day with slack tide. Approach at about 15 degrees, port side to in order to maximise the swing effect when you go astern. Stop engine about three boats' lengths out and when you have half a length to go put the engine astern. This should do two things. First, it should take the way off your boat and second it should kick your stern in. **Fig 16.7**

Conditions are rarely as ideal as that. If there is a dominant headwind approach at a slightly wider angle, say twenty degrees, and put the wheel over to port. This will stop the bow from coming up into the wind. Again stop and go astern at the last moment, but keep port helm on until you have a warp over the side. **Fig 16.8**

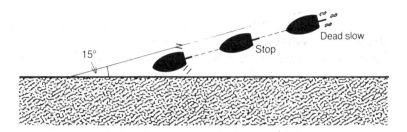

Fig 16.7   Approaching a berth

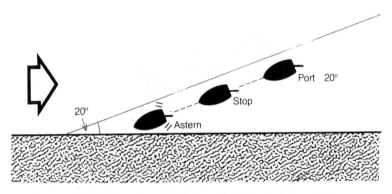

Fig 16.8   Approaching a berth (wind ahead)

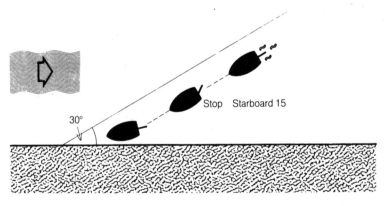

Fig 16.9   Approaching a berth against the stream

With stream dominant you need a bigger angle still. The danger here is that the stream may push you downstream to a berth which you do not want or which may already be occupied. This time put your helm over to starboard to keep your bow out. **Fig 16.9**

In **Fig 16.10** we have an onshore wind and again we have to prevent it from pushing us on to the jetty too early. Approach at an angle of about 30 degrees. Stop and as you approach the jetty go astern. The purpose of sternway here is to prevent you from being hammered on to the jetty.

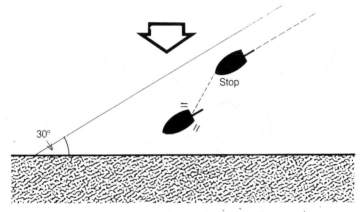

Fig 16.10 Approaching a berth (onshore wind)

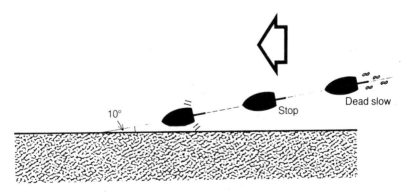

Fig 16.11 Approaching a berth (wind astern)

With wind astern the main object is to prevent the wind from wedging between the boat and the jetty and forcing the stern out. Approach very fine indeed (about 10 degrees) and dead slow. Stop earlier than normal and go astern for longer. **Fig 16.11**

Obviously you cannot always go port side to, so what happens when we berth in a situation where sternway is going to kick the stern out instead of in? In this case approach as normal, but as soon as possible get a spring ashore, either to a helper ashore or to a crew member who has jumped. As soon as it is made fast put the helm hard over to port and go slow ahead. Since the spring prevents any headway, the stern must swing in. **Fig 16.12**

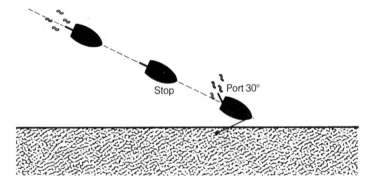

Stop   Port 30°

Fig 16.12   Approaching a berth (starboard side to)

Clearing A Berth

Very often conditions are so calm that one can let everything go and motor slowly out of a berth, or perhaps give a kick astern using a backspring in order to get the bow out. But with an onshore wind this won't work. However, the following routine will. **Fig 16.13**

1 Let go all except forespring, preferably rigged as a slip.
2 Port 15. Slow ahead.
3 Stop engine. Let go spring.
4 Midships. Half astern.

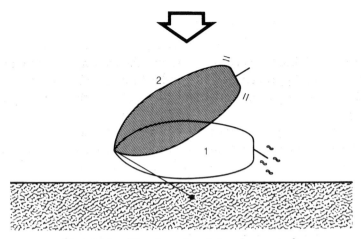

Fig 16.13   Clearing a berth (onshore wind)

With an offshore wind the situation is akin to the first example. Keep two breastropes only, rigged as slips. Put the helm over to leeward, let go and steam off slow ahead. **Fig 16.14**

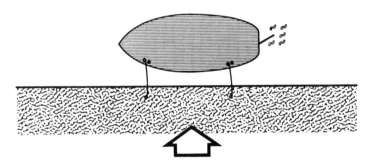

Fig 16.14   Clearing a berth (offshore wind)

Twin Screw Vessels

Apart from being more powerful and having most systems duplicated, twin-engined vessels have one great advantage over single-screw; they don't have any paddle effect. Of course they could, if both screws turned the same way, but twin

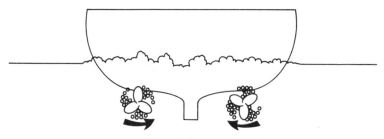

Fig 16.15   Twin screw vessel

screws turn in opposite directions thus cancelling out the pull of the lower blade. **Fig 16.15** Everything is much easier with twin screws, except paying the fuel bills.

Turning Short Round

1 Starboard 10 (you don't need much wheel). Stop starboard.
2 Slow astern starboard.
3 Stop starboard. Slow ahead starboard.

And that's all there is to it. The port engine has been running slow or half ahead all the time. **Fig 16.16**

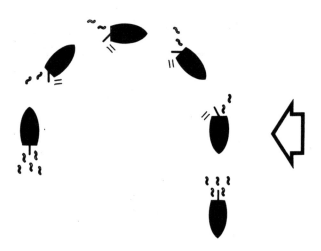

Fig 16.16   Turning short around, twin screw (wind abeam)

Berthing against the Stream

Come in at a fairly tight angle and stop both engines. Close in until practically stationary and put your outer engine astern. It should kick your stern in nicely. **Fig 16.17**

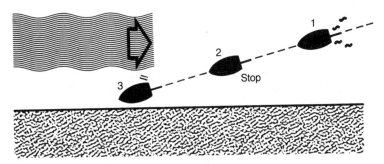

Fig 16.17 Twin screw vessel approaching a berth

Berthing with the wind astern is not quite so straightforward. The important thing, once again, is to prevent the wind from wedging your stern out, so lie parallel with the jetty half a boat's length out and give the outer engine a kick astern. This should bring the stern in, and with that wind there is little chance of your bow being blown out while you get your warps organised. **Fig 16.18**

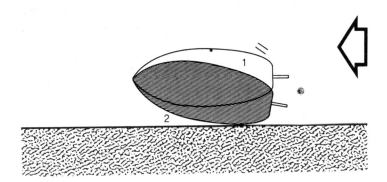

Fig 16.18 Twin screw vessel berthing wind astern

Clearing a Berth: Twin Screw

The only time this might be a problem for a twin-screw vessel is either with an onshore wind or with a strong stream bearing down from astern. In **Fig 16.19** we have the former situation.

1 Reduce warps to one forespring.
2 Put the helm over to leeward. (15 degrees.)
3 Go ahead on the windward engine and astern on leeward.

The spring checks headway and the stern should swing out. A word of warning however. It is risky to use the inboard screw when close up against a jetty unless it is sited well inboard. It would, for instance, be inadvisable to go ahead on the inboard screw whilst using a backspring to swing the bow out. It could be damaged on the harbour wall.

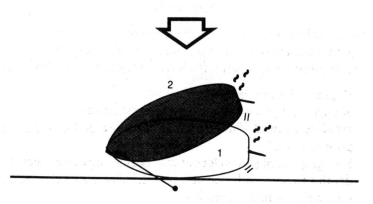

Fig 16.19   Twin screw vessel clearing a berth (onshore wind)

When slipping with the stream, once more reduce warps to a forespring. You might have to be a bit nippy because the current could swing your stern out before you are ready, so make your penultimate warp an after breast rigged as a slip. Let it go and with only the forespring go ahead on the outer engine. With no headway the stern will swing out and you let go the forespring. To get into the fairway, stop the outer

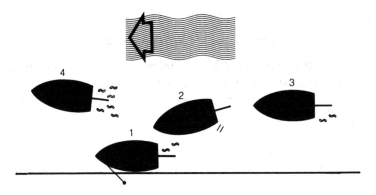

Fig 16.20   Twin screw vessel clearing a berth (stream astern)

engine and go astern on the inboard. Having dropped back to point 3 in the diagram, go ahead, first on inner and then on both engines. **Fig 16.20.**

Anchoring

There is always a time when there is no berth or buoy for you so you must anchor. Get your largest scale chart of the area and decide on your anchorage having regard to the following:

Depth (at highest point of the tide).
Shelter (and you want to be out of the fairway.)
Holding ground (Fine sand and mud are good, rocks are poor holding.)
Swinging room (If you let out cable equalling three times the depth of water, will you still be safe? Draw a circle, radius your cable + boat's length.)
Spoil ground (You don't want to foul your anchor.)

Having settled on a spot that fulfils all these conditions, decide on a course to run in on, preferably one with a transit on which you can steer. Then look for a conspicuous object on the beam which you can use as a reference bearing.

In **Fig 16.21** it is decided to anchor in 4 metres (at datum) in line with the right-hand edge of a large conspicuous house in transit with a church tower. Its bearing is 272° M. Our beam reference is the light on the end of the jetty.

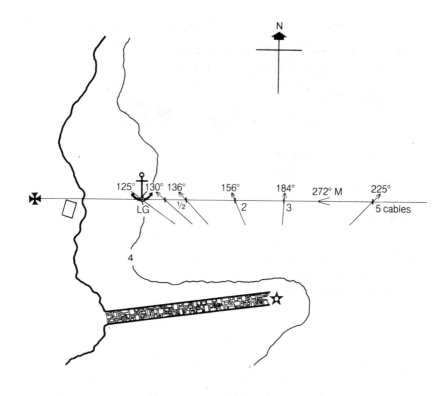

Fig 16.21 Anchor run

Draw the transit back from the church.

Mark your let go point with an anchor.

Using dividers, mark off ½, 1, 2, 3, and 5 cables from your let go point.

From each of these marks read off the bearing of the light.

Record the bearings in your notebook, and hand it to a reliable crew member with the hand-bearing compass.

Come round to 272° well to seaward of the 5 cable mark and line up on the transit. Leeway may cause you to be swung off. If the church starts to bear 269° you are being swung to the north and you must steer 'lower' still (say 265°) to get back on to the transit. If the bearing reads 275° you are being swung south and you must steer 'high', perhaps 280°, to return to your planned track.

When the light bears 225° your crewman will call 'Five cables', and so on until at 136° one cable is reached. Slow right down and at 130° he will call 'Half a cable' when you might stop your engine. When the bearing is 125° he will call 'Let go' and hopefully proceed to do so.

How much cable? As mentioned above, at least three times the greatest depth of water likely in that spot. Datum is 4 metres and supposing that height of tide is 5 metres, then we must allow for 9 metres. Three times nine is twenty-seven, so we must let out at least 27 metres or about 90 feet which is one shackle.

If there is much stream or if you are expecting the wind to freshen, then five times would be more prudent, ie 45 metres or 1.5 shackles. Before doing this, however, take your compasses and go to the scale which is always shown on large-scale plans. Stretch out to 45 metres, add your boat's length, say 8 metres, and describe a swinging circle around your anchorage. Are you clear of all dangers? If not, think again. (Don't worry about other craft unless they happen to be moored fore and aft. They will swing with you.) **Figs 16.22 and 16.23**

If there is danger of swinging against something then you must anchor fore and aft. Having anchored, put the engine slow astern and drop back until you can lower your kedge

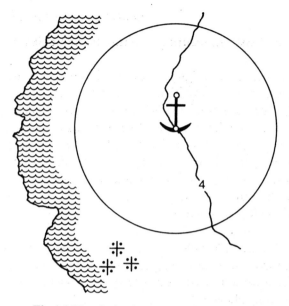

Fig 16.22 Swinging room, three times depth

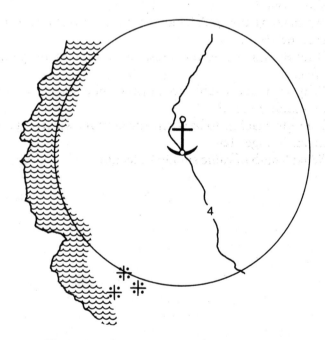

Fig 16.23 Swinging room, five times depth

anchor over the stern, securing it to the strongest cleat or bitts you have.

When you come to weigh anchor which one will you take in first? Answer, the one to which you are not lying.

Now that you are safely in your haven it is my hope that you have found this book useful. It is by no means exhaustive and much reading and more practice is vital before venturing afar. When you do you will be partaking in one of the few activities possible today where skill, knowledge, judgement, nerve and concentration are essential to your survival. May you be successful in all your voyages.

EXERCISE THIRTY-SIX
(a) What three forces act on a power-driven vessel?
(b) In which direction will a single-screw vessel tend to yaw when going ahead?
(c) In which direction should you turn a power-driven vessel short round?
(d) When going alongside a jetty why is it preferable to go port side to?
(e) Why does a twin-screw vessel have little or no paddle-effect?
(f) Why is it inadvisable to use the inboard screw when alongside a berth?
(g) Name five factors to be taken into account when preparing an anchorage plan.
(h) What length of cable should be let go?

17 Chart Run Exercises

CHART RUN FIVE Chart 5051
0700 Castle Treveen Pt (2.3 miles west of Tater du Lt) bore 340° range 2.7 miles. Course 090° speed 6 knots.
0800 Tater du bore 296°, St Michael's Mt 346°, Porthleven Lt 047°. Fix.
A/C to clear Lizard Pt Lt by 2 miles.
1000 Lizard Pt bore 034°. A/C 090°.
1030 Lizard Lt bore 317°. Fix.

Questions
1 What is the Lat and Long of your DR position at 0800?
2 What is marked on the chart nearest to your 0800 Fix?
3 What has been the set and rate of the tidal stream since 0700? (the difference between your DR position and your Fix).
4 What was the course to steer at 0800?
5 What was the ETA Lizard Lt abeam?
6 What is your Lat and Long at 1030?

CHART RUN SIX Chart 5051
You are anchored 1 mile due south of St Michael's Mt
0830 Weighed anchor and set a course to clear Predannack Hd (approx three miles NNW of Lizard Lt) by 1 mile. Your speed 6 knots.

Questions
1 What is your course to steer at 0830?
2 What will be your DR position at 1000?
3 Will you clear the Boa?

1000 Predannack Hd bore 050°. Lion Rock (1.2 miles NW of Lizard Lt) bore 099°.

4 What has been the set and rate of the tide since 0830? (The difference between your DR position and your Fix. Remember to divide by 1.5).

5 Will you still clear the Boa?

6 What course should you now steer to leave the Boa on your port beam?

1030 Lizard Lt bore 062°. A/C 090°.
1100 Lizard Lt bore 343°.

7 What is the Lat and Long of your 1100 Fix?

CHART RUN SEVEN Chart 5067
1000 Dodman Pt bore 350° range 4 miles. Fix.
Set course to clear Lizard Pt Lt by 2 miles. Your speed 12 knots.

Questions
  1 What is the course to steer at 1000?
  2 What is ETA Lizard Lt abeam?

1130 Lizard Lt bore 295°. A/C 270°.
1200 Lizard Lt bore 041°. Fix.

  3 What is Lat and Long of the 1200 Fix?

1200 A/C to leave Runnelstone Buoy 5 cables on starboard beam.

  4 What is the course to steer at 1200?
  5 What is ETA Runnelstone abeam?

1300 Castle Treveen Pt bore 300°. Cudden Pt 046°. St Clement's Isle 016°. Fix.

  6 What depth of water is shown nearest to your 1300 Fix?
  7 What is NEW course to steer to stay 5 cables off the Runnelstone buoy?

1320 Gwennap Hd abeam, range 1 mile. Set a course for Peninnis Head (Isles of Scilly).

  8 What is your course to steer at 1320?
  9 What is ETA Peninnis Head?

1330 Wolf Rock Lt bore 215°.
1400 Wolf Rock Lt bore 125°. Fix.

10 What is Lat and Long of 1400 Fix?

CHART RUN EIGHT Chart 5067
0800 Bishop Rock Lt bore 340°, range 5 miles. Fix.
Set a course to pass 1 mile south of the Wolf Rock allowing for a tide setting 230° at 2 knots. Your speed 10 knots.
1015 Wolf Rock bore 020°. Tide is now 340°, 2 knots.
1030 Wolf Rock bore 320°. Fix.
Set a course to clear Lizard Pt Lt by 2 miles. Tide slack.
1230 Lizard Lt bore 000°. A/C 090°. Tide setting 060° at 2 knots.
1300 Lizard Lt bore 288°. Fix.

Questions
1 What is the course to steer at 0800?
2 What is the ETA Wolf Rock abeam?
3 What is the Lat and Long of the 1030 Fix?
4 What is the course to steer at 1030?
5 What is ETA Lizard Lt abeam?
6 What is your Lat and Long at 1300?

CHART RUN NINE Chart 5051
2100 You are in position 49° 52' N, 05° 43.1' W
2200 You are in position 49° 54.1' N, 05° 34.3' W

Questions
1 What has been your course and speed made good since 2100?
2 What should your position be at 2230?

2230 Tater du Lt bore 328°. Lizard Lt bore 071°. Fix.

3 What depth of water is on the chart nearest your 2230 Fix?

2230 You alter course to render assistance to a fishing vessel in distress, position 270° Mullion Island, range 1.5 miles. Your speed is 6 knots. You discover that there is a tide setting 080°, 2 knots.

4 What is your course to steer for the MFV?

5 What is your ETA at the MFV?

CHART RUN TEN Chart 5067

0900 Trevose Hd Lt bore 100° range 4 miles. Fix.

Set course to clear Pendeen Lt by 2 miles allowing for a tide setting 337°, rate 1 knot. Your speed 8 knots.

1130 Godrevy Island Lt bore 150°. Tide now 240° 1 knot.

1200 Godrevy Island Lt bore 108°. Fix.

A/C to clear Pendeen Lt by 2 miles. Tide slack.

1400 Cape Cornwall bore 093°

Wolf Rock Lt bore 173°

Longships Lt bore 135°. Fix.

A/C to leave Peninnis Hd Lt (Isles of Scilly) 1 mile to starboard.

Questions

1 What is course to steer at 0900?

2 If tide remained constant what would your ETA Pendeen abeam be?

3 What is the course to steer at 1200?

4 What is your new ETA off Pendeen?

5 What is the Lat and Long of your 1400 Fix?

6 What is course to steer at 1400?

7 What is ETA Peninnis abeam?

CHART RUN ELEVEN Chart 5051

A vessel steering 100° C in Mount's Bay wishes to check Deviation on that heading. At 1200 the skipper notices Cudden Pt and a conspicuous water tower coming into transit. At the moment of transit he observes their bearing by compass to be 014° C, Variation is 8° W.

   On the return passage, steering 280° C, he re-crosses the transit line when the bearing of the two objects is 002° C.

Question

What is the deviation on these two headings?

CHART RUN TWELVE Chart 5051. Variation 8° W

For Runs Twelve and Thirteen use the Deviation card overleaf.

Deviation Table for Chart Runs 12 and 13

Compass	Dev	Magnetic
000°	2° W	358°
020°	5° W	015°
040°	9° W	031°
060°	12° W	048°
080°	14° W	066°
100°	12° W	088°
120°	10° W	110°
140°	8° W	132°
160°	5° W	155°
180°	0°	180°
200°	5° E	205°
220°	9° E	229°
240°	12° E	252°
260°	14° E	274°
280°	12° E	292°
300°	10° E	310°
320°	7° E	327°
340°	3° E	343°

You are steering 012° C at 10 knots.

1400 Wolf Rock Lt bore 208½° C range 1.5 miles.

Set a course to clear Tater du Lt by 1 mile allowing for a tide setting 180° T rate 2 knots.

1500 Castle Treveen point bore 328° C. Tater du Lt bore 022° C. Fix.

When Low Lee buoy bore 022° C. A/C 000° T.

Questions

1 What is the course to steer at 1400?

2 What is the ETA Tater du Lt abeam?

3 What is the Lat and Long of your 1500 Fix?

4 What time will Low Lee buoy bear 022° C?

5 What is the Compass course to steer for Low Lee buoy?
   Tide slack.

CHART RUN THIRTEEN Chart 5051. Variation 8° W
An ominous number for our last Run!
You are steering 200° C at 5 knots.
1000 St Anthony Hd Lt bore 306° C, Manacles Pt CG bore
227° C. Fix.
1140 Black Head bore 292° C. A/C 242° T.
1250 Lizard Pt Lt bore 310° C. A/C 270° T.
1350 Lizard Pt Lt bore 030° C. Tide setting 100° T at 1 knot.

Questions

1 What is Lat and Long of the 1000 Fix?
2 What is the compass course to steer at 1140?
3 What is the compass course to steer at 1250?
4 What is the Lat and Long of your 1350 Fix?

Answers

CHAPTER ONE

page 14   **Exercise One**

(a) Pilotage is navigation within sight of terrestial objects.
(b) Because it lacks the latest corrections.
(c) They should be non-ferrous, non-corrodable, heavy and have a span of about ten inches (25 cms).
(d) Its hardness makes it difficult to erase.
(e) It is coloured.
(f) Fathoms, metres and feet.
(g) Soundings are taken from Chart Datum, heights from MHWS.
(h) The dangerous wreck is surrounded by a dotted line.
(i) The one which always shows does not have a line beneath its height.
(j) Admiralty Publication 5011, Symbols and Abbreviations.
(k)

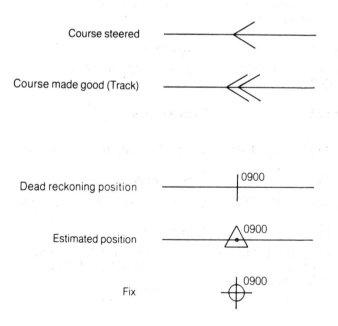

Course steered	
Course made good (Track)	
Dead reckoning position	0900
Estimated position	0900
Fix	0900

CHAPTER TWO

page 16   Exercise Two

(a) 50° 02.8′ N   (d) 49° 58′ N
(b) 50° 07′ N   (e) 49° 57′ N
(c) 50° 08.4′ N

page 17   Exercise Three

(a) 05° 44.7′ W   (d) 06° 04′ W
(b) 05° 40′ W   (e) 06° 26.6′ W
(c) 05° 02′ W

page 18   Exercise Four

(a) Quies Rock   (c) Bawden Rock
(b) Medusa Rock   (d) The Boa

CHAPTER THREE

page 19   Exercise Five

(a) 13.6 miles   (d) 23.4 miles
(b) 3.4 miles   (e) 16.6 miles
(c) 7.7 miles

page 21   Exercise Six

(a) 045°   (b) NW   (c) SSW   (d) NNW
(e) 067½°

page 22   Exercise Seven

(a) 180°   (b) 135°   (c) 270°   (d) 000°   (e) 047°

page 22   Exercise Eight

(a) 246°   (b) 281°   (c) 088°   (d) 054°   (e) 340°

page 23   Exercise Nine

(a) Longships Lt Ho   (b) Seven Stones L Vessel
(c) The Boa   (d) Wolf Rock Lt Ho
(e) Yellow Buoy

CHAPTER FOUR

page 27   Exercise Ten

(a) 5 knots   (b) 5 knots   (c) 2.75 knots   (d) 3.6 knots
(e) 5 knots   (f) 12 miles   (g)21 miles   (h)10 miles
(i) 7.5 miles   (j) 33.75

page 28   Exercise Eleven

(a) Seven Stones LV
(b) 170° 21.8 knots
(c) 293° 10 knots

CHAPTER FIVE

page 31   Exercise Twelve

(a) Wolf Rock   (b) Yellow Buoy   (c) Medusa Rock
(d) Bann Shoal   (e) Wreck in 29 fathoms

page 33   Exercise Thirteen

(a) 50° 05.6′ N   05° 46.8′ W
(b) 50° 02.9′ N   05° 50.1′ W
(c) 50° 02′ N   05° 30.5′ W

page 35   Exercise Fourteen

(a) 50° 19.6′ N   05° 24′ W
(b) 49° 55.8′ N   05° 10′ W
(c) 49° 59′ N   05° 53.3′ W

page 41   Chart Run One

1 236°   2 198°   3 127°   4 071°   5 008°   6 1515

CHAPTER SIX

page 46   Exercise Fifteen

(a) Fixed Green
(b) A light which flashes in groups of three every 30 seconds
(c) An orange light which eclipses every 12 seconds
(d) A red light which eclipses three times every 12 seconds
(e) A red light showing equal periods of light and darkness
(f) A red light flashing more than fifty times a minute

(g) An orange light which flashes more than eighty times a minute
(h) A light showing equal periods of light and dark every ten seconds showing white in one or more sectors and red in one or more sectors
(i) A light which alternately flashes white and red every 30 seconds
(j) A light which eclipses every 20 seconds showing white in one or more sectors and red in one or more sectors

page 52   **Exercise Sixteen**

(a) Can-shaped, red, shows a red light
(b) Cone-shaped, green, shows a green light
(c) Leave it on your port hand
(d)   Q Fl (3)

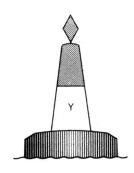

(e) Turn to port

CHAPTER SEVEN

page 58   **Exercise Seventeen**

A   Trawler right ahead
B   Buoy Green 25 (Fine on starboard bow)
C   Rock Green 55 (On starboard bow)
D   Lighthouse Green 100 (starboard beam)
E   Yacht Green 140 (starboard quarter)
F   Container ship dead astern
G   Warship Red 130 (port quarter)
H   Light Vessel Red 90 (port beam)
I   Sailboard Red 70
J   Oil rig Red 40 (port bow)

Exercise Eighteen

A   Alter course to starboard
B   Alter course to starboard
C   Alter course to starboard
D   Maintain course and speed (but be prepared to give way)
E   Maintain course and speed
F   Maintain course and speed
G   Maintain course and speed
H   Maintain course and speed (with caution)
I   Maintain course and speed (with caution)

CHAPTER EIGHT

page 73   **Exercise Nineteen**

1 (a) Vessel at anchor (b) Fishing vessel under 20 metres (c) Sailing vessel under auxiliary power (d) A vessel constrained by her draught (e) Tug with tow over 200 metres
2 A cone point downward
3 She is aground
4 (a) Fishing vessel other than a trawler (b) Vessel trawling (c) A vessel constrained by her draught
5 The lights of a pilot vessel
6 An air-cushion vessel in the non-displacement mode (hovercraft)
7 (a) I am operating astern propulsion
   (b) I am turning to starboard
   (c) I intend to overtake you on your starboard side
8 (a) A power-driven vessel under way and making way
   (b) A vessel manoeuvring with difficulty
   (c) A vessel under 100 metres at anchor
9 The vessel at anchor is over 100 metres in length
10 She intends to overtake you on your port side. You make — . — .

CHAPTER NINE

page 77   **Exercise Twenty**

(a) Mayday (b) Red parachute flare (c) Orange smoke
(d) Sounding of VICTOR or continuous sound of a siren
(e) Raising and lowering of arms

page 77   **Chart Run Two**

1 49° 56.4' N   05° 19.4' W
2 To avoid the Boa
3 49° 54.7' N   05° 09.5' W
4 025°
5 1245
6 000°
7 1702

CHAPTER TEN

page 86   **Exercise Twenty-one**

1 (a) 157°   (b) 0906
2 (a) 078°   (b) 1319
3 (a) 059°   (b) 1958
4 Full and new moon
5 24 hours and 50 minutes
6 All Chart Datum except (c) which is MHWS
7 A MHWS   B MHWN   C MLWN
   D MLWS   E Datum
8 6.5 metres   9 No Clearance. Almost 4 metres too short

page 92   **Exercise Twenty-two**

(a) 49° 56.1' N   05° 05.7' W   (b) 50° 05.7' N   04° 58.4' W
(c) 46 metres

page 93   **Chart Run Three**

1 49° 55' N 06° 05.5' W   2 086°   3 0133   4 Epson Shoal
5 081°   6 0506   7 49° 55.9' N 05° 09' W

CHAPTER ELEVEN

page 99   **Exercise Twenty-three**

(a) The earth's magnetic field
(b) Metal masses in the boat
(c) The difference between True North and Compass North
(d) Deviation is b, Easterly
(e) Deviation is f, Easterly. (In both cases the compass was to the
     east of Mag)
(f) By swinging the vessel through 360 degrees and noting the

difference between a known magnetic bearing and the compass
bearing on various headings
(g) Magnetic and Compass
(h) Variation and Deviation

page 101   Exercise Twenty-four

(a) 251° C   (b) 230° C

page 101   Exercise Twenty-five

(a) 113° C   (b) 300° C

page 101   Exercise Twenty-six

(a) 10° E   197° C   (b) 082° T   072° T
(c) 097° M   095° C   (d) 070° T   068½° M

page 102   Exercise Twenty-seven

(a) 202° M   196° T   (b) 2° E   9° W
(c) 348° C   359½ T   (d) 4° E   217½° T

page 102   Exercise Twenty-eight

(a) 015° T   (b) 159° T   (c) 280° T
(d) 094° C   (e) 250° C   (f) 312° C

page 103   Chart Run Four

1 50° 3.8' N   05° 26.6' W
2 144° C
3 50° 00' N   05° 22.7' W
4 49° 55.7' N   05° 09' W

CHAPTER TWELVE

page 108   Exercise Twenty-nine

(a) 200 kHz
(b) 0555, 1355, 1750 and 0033
(c) Barometric pressure
(d) 6–12 hours
(e) Storm, 48 knots
(f) Less than 2 Kms
(g) Radio 4 at Closedown
(h) Ch 16, Ch 12 and 14, Ch 67

(i) MIKE OSCAR ROMEO LIMA ALFA INDIA X-RAY
(j) Emergency, Distress, Navigational warning

page 116   Exercise Thirty

(a) I have a diver down, keep well clear
(b) I am manoeuvring with difficulty
(c) (i) AA AA AA (ii) AR (iii) T

CHAPTER THIRTEEN

page 125   Exercise Thirty-one

(a) CPA is 3.3 miles at 1115
(b) CPA is 5 miles at 1322
(c) CPA is 2.8 miles at 0402

page 133   Exercise Thirty-two

(a) Wreck in 45 metres
(b) 50° 05′ N   05° 00′ W
(c) 50° 08.6′ N 04° 51′ W

page 135   Exercise Thirty-three

(a) 280° 20.8 knots   (b) 001° 15.4 knots   (c) 235° 7.2 knots

CHAPTER FOURTEEN

page 140   Exercise Thirty-four

1 (a) 301.1 kHz NF (— . .. — .)
  (b) 310.3 kHz DU (— .. .. —)
  (c) 312.6 kHz NB (— . — ...)
  (d) 294.2 kHz GJ (— — . . — — —)
2 Quadrantal errors, distance from beacon, night effect, coast refraction
3 Decca uses a chain of beacons, while Satnavs utilise satellites in Earth orbit
4 They make a considerable drain on power sources. Most require special charts
5 They would be out of range. (10 miles only)

CHAPTER FIFTEEN

page 149   Exercise Thirty-five

(a) Complete in daylight (1st time), fair tide, 1–2 miles offshore, recent weather report and inform Coastguard
(b) Time, log reading, course steered, wind direction and force, pressure
(c) At first light, so that identification of buoys and leading lights is easier and more accurate
(d) You will stay in deep water avoiding outlying rocks and shoals until the last moment

CHAPTER SIXTEEN

page 167   Exercise Thirty-six
(a) Thrust, rudder and paddle effect
(b) To port
(c) To starboard so that the turn is tightened as you go astern
(d) So that the stern swings in when you go astern
(e) It is cancelled by the 'handing' of the screws
(f) If could easily foul the jetty
(g) Depth, shelter, holding ground, swinging room, spoil ground
(h) Three times the depth of water. Five times in strong stream

CHAPTER SEVENTEEN

page 168   Chart Run Five

1 49° 59.8′ N   05° 27.3′ W
2 The 50 metre contour
3 049° 1 knot
4 120°
5 0932
6 49° 55.6′ N   05° 09.4′ W

page 168   Chart Run Six

1 132°
2 49° 59.8′ N   05° 18.3′ W
3 Yes
4 182° 0.8 knots
5 No
6 160°
7 49° 56′ N   05° 11.5′ W

page 169   **Chart Run Seven**

1 229°
2 1140
3 49° 55.7′ N   05° 14.8′ W
4 287°
5 1327
6 29 fathoms
7 267°
8 254°
9 1526
10 49° 58.8′ N   05° 53.2′ W

page 170   **Chart Run Eight**

1 067°
2 1104
3 49° 55.3′ N   05° 46.6′ W
4 090°
5 1257
6 49° 55.7′ N   05° 03′ W

page 170   **Chart Run Nine**

1 070° 6.1 knots
2 49° 55.1′ N   05° 30′ W
3 66 metres
4 027°
5 2336

page 171   **Chart Run Ten**

1 218°
2 1315
3 229°
4 1259
5 50° 07.6′ N   05° 50.5′ W
6 230°
7 1647

page 171   **Chart Run Eleven**
1 6° W   2 6° E

page 171   **Chart Run Twelve**

1 077° C
2 1505

3 50° 00.7′ N   05° 34.6′ W
4 1517
5 011½° C

page 173   **Chart Run Thirteen**

1 50° 07′ N   04° 57.4′ W
2 238° C
3 265° C
4 49° 54.6′ N   05° 13.3′ W

Further Reading

An Introduction to Coastal Navigation ESL Bristol
Lt Cdr Rantzen *Handling Small Boats under Power* Barrie & Jenkins
K. Wilkes *Practical Yacht Navigation* Nautical Publications
Capt W. Moss *Radar Watchkeeping* Maritime Press
Capt Riley et al *Stanford's Sailing Companion* Stanford Maritime (1980)
The Sailor's Handbook The Essential Sailing Manual David & Charles (1983)
R. Willet *Starting Cruising* David & Charles (1983)
Conrad Dixon *Start to Navigate* Adlard Coles (1983)

Acknowledgements

The author wishes to thank Sir Peter Johnson of Nautical Books for permission to use the Mike Peyton inspired cartoon, from *They Call It Sailing,* and his shipmate, John Williams, who checked all the exercises and answers.

Index of Line Illustrations

Index